Robert Mackenzie Watson

Queensland Transcontinental Railway

Field notes and reports with maps showing positions of various camps

Robert Mackenzie Watson

Queensland Transcontinental Railway
Field notes and reports with maps showing positions of various camps

ISBN/EAN: 9783337324490

Printed in Europe, USA, Canada, Australia, Japan

Cover: Foto ©berggeist007 / pixelio.de

More available books at **www.hansebooks.com**

QUEENSLAND

Transcontinental Railway.

FIELD NOTES AND REPORTS

WITH

MAP SHOWING POSITIONS OF VARIOUS CAMPS.

By ROBT. WATSON, M. Inst. C.E.

1881.

Melbourne:
W. H. WILLIAMS, PRINTER, 83 QUEEN STREET.

1883.

PREFACE.

THE object of this expedition was to obtain information for the Queensland Government as to the practicability of constructing a Transcontinental Railway from Roma to Point Parker, Gulf of Carpentaria, on what is called the Land Grant System, as adopted in Canada and other parts of the British Dominions.

These Field Notes were not intended for publication, but some of my friends have persuaded me that they might be interesting and, perhaps, useful. They are for private circulation only.

I have preferred to publish them precisely as they were written down, because a perusal of them in this form, with all imperfections, will carry me back vividly to the surroundings under which they were written.

It will be evident that the Barometical Readings may be disregarded, as there was no fixed barometer with which to correct them, and it is impossible to say whether the variation in the readings was due to atmospheric changes or difference of altitude.

R. W.

MARCH, 1883.

CONTENTS.

———◆———

ERRATA.

On page 31—third line in first paragraph, for " home " read " to me."

  „   fourth „   „   „   for " the kettle for the tea " read "the kettle for my tea."

On page 48—seventh line from bottom, for "and ostensible " read " instead of protracted."

On page 59—seventeenth line from bottom, for "got into" read "got on to."

Queensland Transcontinental Railway.

FIELD NOTES AND REPORTS

OF

ROBERT WATSON, M. INST. C.E.

STARTING FROM ROMA.

FRIDAY, JANUARY 14.—Up at daylight and could not find anybody for half
an hour. Got clear away from the hotel about 6 a.m. It took a long time to
fix the horses, and we did not get away until 7.15. After that there was a
good deal of trouble with the pack-horses bucking off their packs, and the
constable got slightly hurt. The country all the way is very easy. The
Bangeworgarai Creek is crossed a little beyond the entrance lodge of Mount
Abundance Station. The country up to this is very lightly timbered, and
the timber of little or no value. The ground is good all the way, good as
around Roma, and there are indications of plenty of fairly good ballast.
After crossing the creek there is no timber until pretty near to Hodgson, but
the black soil appears to me to be as rich as the Darling Downs. The crossing of the creek will be costly, and there are a few insignificant-looking
water-courses on the way. I think basalt might probably be found underneath the rich black soil. On our way we met a man returning from
Mitchell, where he had taken grapes for sale from Roma ; this looks as if
Mitchell would not grow grapes. We reached Hodgson about 10.15 and
stopped for a time to repair damages, have a cup of tea, and something to eat.
The constable not able to eat anything. I enjoyed some cold corn beef and
tea immensely. Started again at 11.30. About four miles west of Hodgson,
came to Bindango Creek, and the homestead of Bindango Station. Nearly
all the way from Hodgson to this point is rich black soil, little or no timber,
nothing fit for use. In this creek, as in all others, evidently a large quantity
of water comes down sometimes. There is gravel in the bed of the creek,
and lumps of water-worn sandstone some distance on each side. After leaving
the creek the country is more undulating for five or six miles, with, occasionally, a little mild scrub, and now and then a cypresspine. Stone still on
the surface, but of no depth. Then some open level plains, with patches of
timber at some distance off. We passed Mount Abundance itself about a
couple of miles before reaching Muckadilla, about five miles off to the left.
There is only one house, a public, at the crossing of the Muckadilla Creek.
This creek never rises more than six or seven feet, seldom that, and soon goes
down again. We got here at 3 p.m. after a very successful day, the weather
being delicious. I set the barometer at 1000 feet on leaving Roma police
station, and on reaching here it was 1150 feet. I read it frequently on the
way at pinches, but it indicated no difficulties, guessing the distances, of course.

SATURDAY, JANUARY 15.—At 6 a.m. the barometer stood at 1070 ft. I hear that dam-makers get from 11d. to 15d. per cubic yard, finding their own water, the different prices varying with the distance of the water from the work; they find themselves in everything. Difficulty in finding the horses; fourteen were found by seven o'clock, but three were still missing, and among them the constable's. As there were four buggy horses present, I decided to proceed, and leave the other men to find the horses and follow on. The country, all the way to Amby Downs—about ten miles—is open, undulating plains, very sparsely sprinkled with a few scrubby trees and a few bushes. There is ballast to be found all the way; basaltic, I think—hard but very brittle. It occurs in egg-shaped lumps, varying in quality. A lump half as big a man's head, dropped on to something hard from a height of four or five feet, breaks into 2-inch metal: some of it is brown, some grey in colour; many pieces have a sort of black kernel. Met Colonel Halket about four miles from Muccadilla. Got to Amby Downs Hotel about 10.15 a.m. Looking back, could see nothing of the pack-horses, although we could see a long way over the plains. After leaving Amby Downs Hotel, we soon crossed the creek, first in a very zig-zag fashion; then, after a mile or so, again. Here there is some flat, heavy country, and, on the right, a small range of hills, in which I am sure there is ballast. The range is kept on the right for a mile or two, when we enter a smaller range, with more scrub than any I have yet seen. Here is plenty of stone for ballast, and, I think, even for building culverts. The stone is a kind of porphyry. The ranges are so small that they can easily be dealt with: a few iron-bark and bottle trees, but not a blade of grass. This continues for a mile or so; then there is flatter country and equally barren, with iron-bark and Brigalow scrub, also an occasional sheoak. Presently it gradually turns into decent-looking plains for a short distance; but if the grass here (which is abundant) is good, why is it not eaten? About three miles before reaching Mitchell we came to a considerable creek, whose name I could not learn. The country all the way seems poor compared with yesterday's. Immediately before reaching Mitchell the Maranoa River is crossed by a constructed ford, with steep approaches, which are metalled with a fair sample of gravel, suitable for railways. Called at the police office to enquire about the best camping ground, and was told it was near "the dam," about two miles towards Charleville. Drove on, and after great difficulty found a tolerable place on the river near the dam, but where the rest of the party could never find us without a guide. Got some tea, salt-junk, and damper; for, it now being near four o'clock, and having breakfasted at six o'clock, we were hungry. The country along the telegraph line to the dam is flat and easy. Mitchell is a miserable-looking township; three or four publichouses—nasty ones—and the usual proportion of other shops and stores. The remainder of my party turned up all right about 6.30 p.m., pretty tired, and I am tired too. Close to the hotel at Amby Downs is a very small patch of Indian corn, looking very healthy. It was only a very small patch, but sufficient to show what can be done. I was mistaken: grapes can be grown in Mitchell; also peaches, melons, Indian-corn, &c. Mr. Turnbull is the local surveyor.

SUNDAY, JANUARY 16.—In camp all day, busy writing letters, &c. Hear very bad accounts of feed and water towards Charleville. There is feed and water at a place called Sandy Creek, about eighteen miles from here, and we shall have to camp there to-morrow night, only making two-thirds the distance I had intended going, viz., to Moore's Creek, but there is neither feed nor good water there. Mr. Wyatt and Mr. Baynes went into the township and saw the surveyor, who told them that there is plenty of ballast (basalt) all the way from Roma. It seems that what I have called porphyry he calls yellow basalt; it is, I think, a volcanic production. There is also gypsum in thick layers, 12 inches thick, about eight miles off. Must try to see him myself to-morrow morning. A good many ducks about, and some wild hogs. We have camped on the Maranoa River, close to a water-hole, which is the only one for miles, and here the water is bad. I am told that sometimes you may ride along the bed of the river for seventy or eighty miles and not see a drop of water, yet the water channel proper is about 250 feet wide and 30 or 40 feet deep, and the water sometimes extends over the plains for a long distance.

MONDAY, JANUARY 17.—Started at 7.30 a.m., and drove into the township. Called at the police office, and saw Sergeant Byrne and the postmaster, who were both very obliging. Called at the saddler's, and got three saddle pouches, costing 20s. Then drove towards proposed camping place, at Sandy Creek. For the first six miles from Mitchell, the country is fine open plains on either side, as far as the eye can reach; then myall and ironbark scrub for six or seven miles. On entering the scrub there is a pretty steep rise for a mile, but a gradient of 1 in 50 would do it. There is a good sprinkling of ironbark and some box. Sufficient sleepers can be got along the line. Afterwards the country becomes a little more open, and, if anything, a little better, but not of much account, I think. About fifteen miles from Mitchell there is a small hut, and what looks like Cobb's stables. Plenty of ballast all the way. When we reached Sandy Creek there was not a blade of grass to be seen, so the pack-horses and men made for Moore's Creek, leaving a fresh pair of horses for us if we required them; Moore's Creek being about sixteen miles further on. However, we did it, and without changing horses. We got to Moore's Creek at 1.45 p.m., and really the ground looks nearly as bad here. However, camp we must, and we are told there is feed a mile and a half off, so we camped close to the hotel. The country, from about six miles this side of Mitchell, is wretched: nasty brigalow and ironbark scrub, and not a blade of grass all the way, or nearly all the way, to Moore's Creek. There is some difficult country at one place: there is a rise of 300 feet in about three miles, then a fall of 80 feet in about a quarter of a mile, and 35 feet of this is in about three chains. There is ballast all the way or nearly so, and sufficient ironbark and box to supply sleepers for many miles. The bad country I have spoken of extends for about twenty-one miles.

TUESDAY, JANUARY 18.—Horses are in camp in good time. Tried to get some information about the country from the landlord. Started at 7.50 a.m. Had some fresh horses. They went away quietly enough, but about a mile on there was the Devil's own row. However, nobody was hurt. One of the wheelers got frightened, and frightened all the rest; but we got away again without damage. The country for about a mile round Moore's Creek is a little better than quite desolate. Then it becomes bad again for three or four miles: thick Brigalow scrub and not a blade of grass: some stone: then it becomes a little better for about three or four miles, showing indications that sometimes there is grass there; and as evidence we saw two or three small mobs of kangaroos. About seven or eight miles from Moore's Creek the country becomes a little better, not so thickly scrubbed or timbered—an ironbark occasionally in the bad country. From this to Black's Waterholes, about three or four miles further on, the country is better: box timber, and some very similar to the Campaspe Plains before they were cleared, and where excellent wheat is grown now. After leaving Black's Waterholes, the country alternates good and bad: a long stretch of bad, then indications of there having been grass at some time. I think the better portion would grow wheat ; and there are some rising grounds, with red sandy soil, where excellent grapes might be grown. I think this because, on similar red sandy soil near Roma, excellent grapes are grown. It has been a heavy day for the horses. We got to Saddler's Waterholes, all of us, at a little before 3 p.m. I am sure there is ballast procurable everywhere. I do not think a 1 in 50 gradient will require any heavy work. It appears that, all the way from Black's Waterholes to Charleville, water is provided by dams. This could be as well done everywhere.

WEDNESDAY, JANUARY 19.—Made a pretty good start at 7.30 a.m., after a very warm night, and flies troublesome. Horses looking better for being in paddock all night without hobbles. The country all the way is very fair, lightly-timbered, and not unlike Campaspe Plains, but there is so much good land elsewhere which does not require clearing that it must be a long time before it will pay to clear this. Called at hotel and paid for paddocking horses, and at police station, leaving the heavy tent pegs there, as we are not likely to want them. I am told that a few miles to the north there are rich open plains. The country is simple and uninteresting for five or six miles ; then comes a magnificent sheet of water on the right, called the Eurella Dam. From this the country improves, and will, no doubt, sometime or other, be cultivated. We passed Leadbeater's, a coach stage, and soon after began to follow down the nicest creek I have yet seen, Hamburg Creek, a succession of splendid waterholes for miles and miles. At last we reached Read's, where we expected to find good feed, but, alas ! it was like all the rest. Read's is a small station, not a publichouse. Horses are all getting knocked up ; they cannot live without food. Ballast all the way : a few ironbark, box and sandalwood trees, the latter of no account. On reaching our camping ground, about a mile beyond Read's, we found excellent water, but very little feed. The heat has been intense, and it continued until midnight—the flies dreadful and the ants bad. Horses very much done up. Some ducks were got in the evening. I find the dams I have mentioned have been made by government.

THURSDAY, JANUARY 20.—Made a good start at 7.20 a.m., having seen all the horses right first. When about two miles on the road, all the pack-horses overtook us but one, which the men said had broken away and was lost to sight. The constable, groom, and black boy were after him. We passed a lot of water-holes on our way, and there are some tolerably open plains, with not much timber or scrub. On our way we got a turkey. It appears that the Hamburgh Creek, on which we camped, and the Clara, join a little further down, and form the Angelalla Creek, which falls into the Warrego at Maugalore. About six or seven miles from camp, say thirty-five miles from Charleville, we came to the first range we have had to cross. I think its summit is about 230 feet above the plain, and in the last quarter of a mile before reaching the summit the rise is about 130 feet. The formation is volcanic, and the stone, which is called yellow basalt, is rather soft, and there is plenty of it. There is no useful timber, and the country is very poor ; no grass, only wretched scrub. The lowest point, I fancy, has not been selected for the crossing of the road and telegraph. Survey of this should be made. This is called the Angelalla Range. After reaching the summit the road is level for a mile or two, and, although there is not much grass at present, there are indications of plenty sometimes. The road is stony for many miles, and gradually falls. In about three or four miles miles it has got down about 130 feet, and then a level plain continues to the dam, twelve miles from Read's and thirty-two from Charleville. Plenty of water at the dam, and good feed too, some twelve miles further on, we are told ; and there is good feed at one or two places on the way, but everything is wretched at the dam, and the horses will have to go five miles to feed for the night. No sign of the constable, &c., at 1.30 p.m., but at 1.50 all turned up right. On our way we learnt from a traveller that there is much better feed and water on the Nive River than on the Ward River. Sent the horses out to the pasture and soon heard that it was excellent. The coach-groom showed the way.

FRIDAY, JANUARY 21.—Got a pretty good start at 8.45 a.m., considering the horses had to be fetched so far. All the rest of the party ready to start. The country all the way to Charleville is very level. The land is of fairly useful quality, varying as to vegetation, some parts ·pretty thickly scrubbed, some almost open plain. On the whole, it is evident there is abundance of grass sometimes. On nearing Charleville there are some bloodwood trees, and there is a new fence and rail, with cypresspine rails and bloodwood posts, neither very pretty. On arrival at Charleville found there was no feed within seven miles; no horse-shoe nails, and a lot of horses want shoeing. We arrived at noon. Then presently came some of our men, and said two chestnut horses were missing. This is gross carelessness. I did not observe any indication of ballast all the way to-day, at least on the surface; and there is very little timber that is of any value, only an occasional ironbark and curragong. Saw Police Magistrate Oakden and Inspector Thornton. The ants and flies are dreadful. On our arrival at Charleville, about noon, we expected to find plenty of feed, so that our horses could pick up a bit whilst we got the pack-saddles, &c., repaired, and some of the horses shod. To our dismay, there was not a blade of grass to be seen, and there was not a horse-shoe nail in the township. A native police-constable was at once despatched into the bush to look for grass and water. He came back in the afternoon and reported good grass and water at a place called the Clay Pans, about nine miles distant. Decided to let the horses wander about till morning, carefully watched, and then send them out to the feed. At the same time I telegraphed to Roma to have some nails sent by the first coach. They cannot be here before Tuesday night, and if we get the horses shod and the saddles mended by Wednesday night, we may get away on Thursday morning. This is two days later than I intended, but there is no way out of it. Received and sent a lot of telegrams to different people.

SATURDAY, JANUARY 22.—Sent the horses out to the Clay Pans. Heard a most favourable account of the feed, but the water is thick—it may yet be good. Busy all Saturday, writing out draft report, &c. The groom had to go out to watch the horses at night. Telegram from Mr. Lukin; he is most anxious for information about the Grand Trunk Line.

SUNDAY, JANUARY 23.—Saw Inspector Thornton and Mr. M'Farlane about Charleville to Cunamulla and Thargomindah. I got a lot of imformation from them.

MONDAY, JANUARY 24.—Got more information from a Mr. Davis, an intelligent man, who has been with survey parties, but cannot keep from drink. This information will be found in another book, which will be put aside for that line. Wrote my first progress report—Roma to Charleville, and copied my notes about Grand Trunk Line, in case it be authorised that they go to Mr. Lukin. I find Hann will reach Brisbane on 27th, and join me at Tambo or Blackall as soon as possible.

TUESDAY, JANUARY 25.—I got Mr. Wyatt to copy my report and also my field-notes into a book. I got a few letters from Victoria in the evening.

The nails came and the horses were brought in for to-morrow's start; the saddler, too, is getting on all right. We have a thunder-storm every night, but no great fall of rain, only sufficient to lay the dust.

WEDNESDAY, JANUARY 26.—Busy all day about rations, stores, cooking utensils, &c.; and watching the blacksmith and saddler, who both had finished about 4 p.m., so that things seem pretty straight for to-morrow. The horses look fairly well, but there are some sore backs, one very bad one. The cook has been suffering from "Barcoo" for a day or two, but is better. Learnt from Mr. Stanley that the reduced level of Charleville is approximately 927 feet above low water at Brisbane. The township of Charleville is on the east side of the River Warrego. The river overflows its banks at time of flood, and the water has been known to be a foot deep in the principal street.

THURSDAY, JANUARY 27.—The horses were sent to some tolerable feed about three or four miles away, but did not reach the camp until nearly nine, as five of them had gone away for their previous camping ground. However, after paying my bills, determining on my route, and telegraphing to the police at Tambo, we got fairly started altogether at 10.30 a.m. There had been a good drop of rain during the night, and the air was tolerably cool. We crossed the river at once, and proceeded up what is known as the Nive Track, the route being Charleville, Cattle Station (Gowrie), Ellangowan, Nive Downs, Landsdowne, Tambo. For the first six miles the country is open level plain, pretty fair soil, and very lightly timbered; then it becomes more timbered and better soil; evidently plenty of grass sometimes. There is a good sprinkling of bloodwood and blackbutt or Moreton Bay ash, besides a little box, &c. There will be no difficulty in getting sleepers for the whole distance we have come to-day. We followed the road not very far from the River Warrego. The flats extend a long way, and are laced with anabranches. The road is evidently three or four feet under water at times. About twenty-two miles from Charleville there is a curious lot of stone, waterworn, very heavy, and hard; some is basalt; some a conglomerate of cemented gravel, very hard and heavy; some very large boulders. It is not far from the west bank of the Warrego, but there are many anabranches intervening. It is near a sandhill and an old sheep-yard. About three miles further on I noticed a range to the westward, at some distance. I think it must be the west side of the Ward River, because the distance between the two rivers at this point (twenty-five miles from Charleville) is only fifteen miles, as scaled from the new map of Australia. We saw the telegraph line once or twice on our way, so I think it cannot be straight. We found a convenient camping-place about twenty-eight miles from Charleville, pointed out by Constable Montford Bromley, who is accompanying us. Mr. Sandeman passed, on his way to Charleville, just as we were camping. Plenty of good feed and water; plenty of flies too. No doubt, through all the distance we have come to-day, the country is easy and there is plenty of ballast and sleepers; but the line must be kept further from the river than the road is, probably half-way between the Nive and the Ward. It is said the land is better on the Ward, but it is sufficiently good here. Near our camp (No. 8) there are some small but nice open plains, with excellent feed even now. Barometer, set 927 at Charleville, read 880 at 10.30 a.m. to-day, and 1125 at the camping-place at 3.45 p.m., twenty-eight miles from Charleville.

FRIDAY, JANUARY 28.—Barometer at 7 a.m., 1030 ft. We got fairly away altogether at 7.30 a.m., and immediately after passing the cattle station we crossed the Warrego River. As there is no track on the west side, we kept well within sight of the river for a considerable distance, through moderately good country, light-timbered for about six or seven miles. Then, getting further from the river, we came to better country. At about nine miles, came upon some sandhills, and stone similar to what we saw yesterday. Then some open plains, with rich black soil, some salt-bush and tussock grass, but no feed. This continued for about three or four miles, and then came some scrub, sandalwood bushes, but no grass yet. The soil looks good, as if it would grow anything under cultivation. There is a sprinkling of bloodwood, blackbutt, and ironbark, and there are patches of scrub called Dead Finish (an acacia), because it is almost impenetrable. There is water nearly all the way (12 miles), but very green and dirty. The country improves as we go northward, and becomes more open. Although there is very little feed at present, there are evidences of plenty of grass sometimes ; in fact, all the ground looks as if it was ready to burst out into vegetation on the slightest provocation in the shape of rain. We crossed the Yo Yo Creek and Burinda Creek. When about three miles from Ellangowan we began to see much better feed. They have evidently had more rain, and the ground has immediately responded. We found, or were shewn by Constable M'Grath, a capital camping-place, about one mile south from Ellengowan, where there are plenty of turkeys and ducks. Many of the horses are completely knocked up, even some of those which have carried nothing but themselves to-day. We reached the camp (No. 9) at 2.30 p.m.; barometer, 1280 ft. The necessity for crossing the Warrego as we did arose from the fact that, on the western side of the river, the ground for a long distance is heavy and sandy, and subject to floods. I fancy the ground is flooded for a long distance on each side at times. If the line be taken between the Nive and the Ward Rivers, or in the valley of either, it will have to be kept a good distance back, and there is no difficulty. I am confident ballast and timber can be found within reasonable distance all the way, and the land some time or other will all be cultivated. The clearing, however, light as it is, will for some time be an obstacle. I don't think this is always taken sufficiently into consideration. If good land, which can be got at upset price, is worth clearing at £5 per acre, then is not land which nature has cleared, and which is equally good, worth £5 per acre more than the upset price ? On our way I noticed some ranges which I think must be between the Ward and the Nive Rivers, but I do not think they can be of any consequence, as they do not present a pointed appearance. Just as we got camped a heavy thunderstorm came on. We were only just ready for it. It rained heavily for some time, and this was followed by a deliciously cool evening and night.

SATURDAY, JANUARY 29.—Barometer at 8 a.m., 1135 ft. Got away altogether a little before 9 a.m. Splendid feed all the way from the camp to the township of Ellangowan, about a mile. Stopped a short time at the police station. The inspector or sergeant knows the north well, and thinks from what he has heard that there is no doubt whatever about our being able to drive all the way through. Got some repairs done to one of the pack saddles. The country all the way to Burinda Station — Mr. Gordon Sandeman's—is magnificent, and all showing the effects of recent rains. About a month ago there was not a blade of grass ; now there is abundance all the way. The distance to the station from Ellangowan is about twelve miles : the first six lightly-timbered and fine feed ; then comes a most magnificent undulating plain, full of what is called summer-grass, very good feed, large tussocks, and splendid under grass. This plain continues for about two miles, and reaches nearly as far as the eye can see on either side. No timber for a long distance. Then comes undulating plains with a different sort of grass, and small patches of saplings and scrubs—very little. Then, when nearing the homestead, is some fencing. Altogether, it is a magnificent sight. We reached our destination about 12.15 p.m. I was dreadfully sleepy all the way. Obliged to drive to keep myself from falling off the seat. Camped close to Woolshed (Camp No. 10). Barometer, 1270 ft. Called at Burenda Station in the afternoon. Mr. Sandeman not at home, but expected to-night. Saw Mr. Inspector Thornton. He says I have not lost much from coming up the east side of the Warrego from the cattle station, as there is no proper track on the west side of the Nive River, the country being very heavy and sandy, with thick scrub in some places. Returned to Camp No. 10, at the woolshed, to sleep. I noticed to-day, close to Ellengowan, a Chinaman's garden, in which were growing luxuriantly peaches, grapes, melons, cabbages, sweet potatoes, &c. Received telegram from Colonial Secretary, that I might give any information I pleased to Lukin and Buzacott as to my progress.

SUNDAY, JANUARY 30.—Wrote to Mr. James Forrest (Inst. C.E.) in morning, then went to Burenda Station. Mr. Sandeman came home last night, but the Duke of Manchester has not arrived yet. Mr. Sandeman gives the same description of the land between the Ward and the Nive that Mr. Thornton does. There is no difficulty about making the railway along the divide. He says that what I shall see in going from here to Tambo, *via* Nive Downs, will give me a good idea of what the portion I have not seen is like. No difficulty in making a railway, but everybody says, "Whenever will your traffic come?" Mr. Sandeman says their dams cost them about 1s. 7d. per cubic yard, including everything; no punning required, the trampling of the horses and the carts being sufficient. The distance of the dam work from the water regulates the price to a great extent. Whilst at Burenda Station, in afternoon, about 3 p.m., a terrific thunderstorm came on, and lasted for an hour. The rain fell in torrents, and I never saw such lightning or heard such thunder before. One clap scarcely followed the lightning, it accompanied it. It was estimated that about two inches of rain fell. Strange to say, there is no rain-gauge here. The rain which fell last Friday was, I believe, just as heavy, but the ground was dry, and quickly swallowed it up, but now the ground is saturated and the plains are exceptionally heavy. It was with difficulty I got back to the camp in the evening; the horse sunk to his fetlock at every step. A stop is put to our intended start to-morrow. Mr. Sandeman, who is an experienced squatter, does not think much of the Mitchell country, but he looks at everything from a present grazing point of view. He does not deny that eventually, where required, a very large portion of the ground over which I have travelled may be profitably brought under cultivation, but not for many years, nor until the population is greatly increased and more capital brought into use. I am still of opinion that this will not be until the holdings are smaller.

MONDAY, JANUARY 31.—Barometer flitting about in an extraordinary manner. Ground on plains awfully heavy. It is useless to attempt to start. Got some writing done, and, at about 10 a.m., walked to station, thinking to go before lunch with Mr. Sandeman to his dam, but the ground was too heavy. Heard the Duke was expected to-night, but not until late, as he was going to Nive Station, *via* Okewood and Ellengowan, to this place. Mr. Morrisett brought me a telegram from Mr. Hann. It came *via* Tambo and Nive, from Brisbane. He has had an accident, and cannot join me before end of next week at Blackall. Things are looking crooked, but I suppose will come right. In the afternoon I went with Mr. Sandeman to see his garden, about half-a-mile down the creek. It is managed by a Chinaman, who gets twenty-five

shillings per week and his rations, besides the privilege of selling all that is
not wanted for the house. Some men were here engaged making a dam
across the creek in a very primitive style; it will surely wash away at first
flood. The workmen are ordinary station hands, engaged as teamsters, and
when the ground is in such a state as to prevent the teams working, the men
are employed at work of this sort instead of being idle. I am told, and
in fact shewn, that the stone about here is limestone and not basaltic. It
may be that the stone at Mitchell Downs and elsewhere is also not basalt; no
matter, it is all good for ballast. Embedded in the stone here I saw some
fossil shells; one a bivalve about $1\frac{1}{2}$ inches long and $\frac{3}{4}$ inch wide, firmly fixed.
The stone is greenish grey after exposure to the air, but dull blue when
quarried. It fizzes on application of diluted nitro muriatic acid. As there
seemed to be some doubt about the Duke's arrival to-night, I left, and walked
with Mr. Sandeman to the camp about 6 p.m. I am told that the land is very
much better on the Ward River than on the Nive or Warrego; and it may
be worth while to enquire into this when the detail survey is being made, and
before the position of the line is definitely fixed. If the line is taken through
poor land and the Syndicate get nothing but useless land for constructing the
line, this will not bring much traffic. I do not know that the Act defines the
position of the land to be given in lieu of money for the construction. The
distance back from the line is not stated; it could not be, because the quantity
per mile is not stated; it only affirms that no block shall have more than five
miles frontage to the line.

TUESDAY, FEBRUARY 1.—Barometer, 1040 ft. at 7.30 a.m., and it read
1270 ft. at same place (Camp 10) when we arrived at 12.15 p.m. on Saturday.
How it has varied in the meanwhile I don't know. I can tell the difference
in reading at the time of reaching and leaving a camp, but what it does on
its account when I am travelling I cannot say, because I cannot make
arrangements to have the changes on a fixed barometer noted at the same
time. I think there should be a standard barometer and rain-gauge at every
police station or other government office in every township. I find the Duke
arrived at the station in good time last night. Saw His Grace at 9.30. He
is enjoying his trip immensely. He never saw so fine a pastoral country in
all his travels, but cannot see the necessity of a railway from Roma to
Blackall. He thinks it waste, and that the line should go from Emerald to
Blackall instead, as the port at Rockhampton is as good as at Brisbane. Let
us now see:—Brisbane to Roma, 317 miles; Roma to Charleville, 180 miles;
Charleville to Tambo, 120 miles; Tambo to Blackall, 80 miles. 380 miles to
construct, 697 miles to travel. Rockhampton to Emerald, 180 miles;
Emerald to Aramac, 200 miles; Aramac to Blackall, 100 miles. 300 miles
to construct, 480 to travel, in either case, to a shipping port; but the 180
miles from Roma to Charleville will be 180 miles of the Grand Trunk Line,
and I believe the Government is committed to Mitchell, viz., 60 miles. It
also forms part of the connection between Queensland and New South Wales
at Bourke. Doubts are entertained whether or not any English Syndicate will
take up the scheme unless it can be shown pretty clearly that the traffic will
be sufficient to pay the expense of working. I sent the party on in the
morning to camp at the crossing of the river (12 miles), whilst I stopped to
have a chat with His Grace. He seemed pleased with everything, but cannot
see how the railway is going to pay at once; neither can I. About noon I
started with Mr. Sandeman to see his wash-dam, about five miles on my way
to the camp (No. 11) on the Warrego. It is a splendid dam, and must have
cost a lot of money. They have, however, made a mess of the byewashes;
but they are difficult to deal with. Reached camp about 3 p.m., on horseback

B

the last six miles. We passed over magnificent open plains for nearly all the way—then some poor scrubby country until we reached the river. Barometer at Camp 11, at 4 p.m., 1240 ft. In course of conversation, the Duke asked what rent was paid per square mile for squatting leases; and I think the reply was 10s., 15s., and 20s. for the first three years. He then asked what rent it would bear if they could get a twenty-one years' lease. He said, "Would it bear £10 per square mile?" "Oh, dear, no," said Mr. Sandeman. "Would it bear £5?" "Well, scarcely that, but perhaps it would." Now, land is frequently bought for pastoral purposes at £1 per acre, or £640 per square mile; and the interest of this at four per cent. is equal to a rent of £25 12s. per square mile per annum; or, if bought at 10s. per acre, at £12 16s. per square mile per annum. The place at which our camp (No. 11) is fixed is at the crossing of the Warrego, called Caroline, and about twenty-eight miles above the junction of the Nive. There is blackbutt and gum along the banks, and a soft sort of sandstone, not much good, cropping up in large lumps, very brown and soft.

WEDNESDAY, FEBRUARY 2. — Barometer at 7.30 a.m., 1060 ft. The country from the Caroline Crossing of the Warrego to Nive River is very uninteresting. For the first three or four miles alternate open plains, and moderately-timbered plains; then, for about ten miles, what appears to me to be useless scrub, the only redeeming feature being that there are occasionally a few ironbark trees, which may supply some sleepers, and some cypresspine. There is not a blade of grass, and there is no appearance of there ever having been any. The track is pretty good generally for driving over, but we passed through several nasty places called "melon holes," where the waggon taxed the horses to their utmost. These melon holes are low pans, not unlike crab-holes in Victoria; the water settles in them, and they gradually sink lower and lower. About four miles before reaching our camping ground on the Nive, near the station, we entered a paddock, and soon the feed looked better. We found a nice camping ground on the bank of the Nive before crossing. No telegrams; sent one to Barron and to Walsh. At station for an hour in the afternoon with the manager, Mr. Brodie. He says there is no difficulty whatever in running straight from Charleville to Tambo, except want of water. The country here is very beautiful, and there is a good sprinkling of timber, some of it useful. The barometer at 6 p.m., 1180 ft. The flies have not been so vicious at any camp before. This is camp No. 12.

THURSDAY, FEBRUARY 3.—Barometer at 7 a.m., 1005 ft. Got a good start shortly before eight. A little difficulty in crossing the river; heavy sand and fresh horses. We presently came to, and drove for about twelve miles through most magnificent rolling plains, well grassed, and looking really beautiful as far as the eye could reach in all directions. There are a few trees here and there; very few and very small. After about twelve or thirteen miles, we came along the edge of a very lightly-timbered and slightly-undulating country, well grassed, with short, green herbage; and we followed a stream in which there are occasional water-holes. The river appears to be running east or south of east, and our course is west. The country gradually becomes less attractive, and within a few miles is useless, barren sandy scrub, with a good sprinkling of "Dead Finish," and a few iron-bark, blackwood, and blackbutt trees. There is limestone cropping out every-where all the way, in large blocks in some places. We reached our camp (No. 13), near the cattle-station belonging to Nive Down Station. On our way up the river, we rode abreast of some splendid reaches in the river, which I am told is the "Nive." The camping place is a very nice one, with plenty of grass and feed. Just before we started from Camp 12 a young man, who is engaged horsebreaking at the station, came over. It seems he has been several times overland to the Gulf of Carpentaria, and says he never experienced any difficulty. He thinks we shall be ten weeks in reaching the Gulf, as he estimates the distance we have to travel as 1100 miles if we go by way of Cloncurry. We might attempt a shorter cut across the flats, but the floods would stop us. He says we can drive perfectly well all the way, and that there is a track by which the distance is only 800 miles. I cannot understand this. I find we did wrong to turn from the track we were taking this morning towards Landsdowne. We should have saved five miles in distance, and had a better road. However, we have seen some fine country, and have kept near what will probably be the line eventually, besides having a splendid camping ground. Barometer at Camp 13, at 12.45, 1215 ft.

FRIDAY, FEBRUARY 4. Barometer at 7.30 a.m., 1140 ft. Got a good start a little before 8 a.m. For the first twelve miles passed over beautiful park-like downs, with lightly-timbered country on one side. A great lot of well-to-do looking cattle all round. The prettiest sight I have seen for a long time. At about twelve miles, we crossed a fence, and immediately afterwards the telegraph line, whose bearing by compass was N. 55′ W. Then we followed the telegraph line for several miles. The country still good, but perhaps not quite so good as the first twelve miles; a little scrub here and there. The country all appears to be fenced in from where we camped last night to Tambo; and nearly all open plains. The whole distance is said to be about twenty-eight miles; and, in the last twelve miles, there is scarcely a blade of grass; still there are sheep and water here and there. About twenty miles from Camp 13 there is a dam and a good reservoir of water; then there is a succession of waterholes in a tributary of the Barcoo. At Tambo there is no horse-feed. The horses will have to be sent out four or five miles to a police-paddock, and we must try to push on ten or twelve miles to-morrow. Two of the horses are knocked up through what seems to me to be mismanagement. The camping place (No. 14) is a very miserable one, about a mile from the township; not a blade of grass, only dirty water and wood. Telegrams from Barron and Walsh. Mr. Hann cannot join us before to-morrow week, the 12th inst. Barometer at camp at 2 p.m., 1102 ft. All the way we came to-day there is plenty of limestone cropping up, but a great scarcity of timber. There are very few defined water-courses. Tambo is on the Barcoo River, and the distance from it to Blackall is said to be eighty or eighty-five miles, and no feed for the first ten or twelve miles. Sent telegrams to the Colonial Secretary and to Mr. Collier. With reference to following up the dividing ridge between the Ward and Nive, in a tolerably straight line from Charleville to Tambo—this would probably divide the country into two pretty equal parts; and, as the distance between the two rivers varies from nothing to more than thirty miles, for some portion of the way the proposed twelve and a-half miles survey on each side would monopolise both sides of each river, and in some cases it would not reach either. I think it worth consideration whether or not the survey should in all cases stop at the rivers, thus giving all the surveyed blocks a river frontage, and reserving a river frontage on the other side for the unsurveyed blocks. The maps furnished are not by any means accurate; for instance, the recent map of Australia, published in Melbourne, shows the telegraph line from Charleville to Tambo perfectly straight, whereas in fact it is much more like a ram's horn, or the hind-leg of a dog, than an arrow; and the Nive Station is shown about half-way between the two rivers, whereas it is on the Nive. Although the line should select nearly the dividing ridge all the way, there

would be no difficulty in providing water for the engines at intervals either
by pumping from the river or else by dams or tanks near the line.

SATURDAY, FEBRUARY 5.—In camp all forenoon, writing, &c. Walked
to the township. Called at the telegraph office. Instructions to report
from Blackall and all important points. Called at Queensland Bank,
and got cheque cashed. Intended to have started at 2 p.m., and go
on twelve or fifteen miles towards Blackall, where there is good feed; but
as there was some misunderstanding as to my instructions, we have to wait
until to-morrow; and, although to-morrow is Sunday, we must make up then
what we have lost to-day. A smart shower in the afternoon, laying the dust
a little. Telegram from Acting Premier (Mr. Palmer), expressing his annoyance at a paragraph in the *Courier*, which stated that Hann is leader of my
party; that he has taken steps to contradict it, and that Mr. Hann fully
understands his position of bush-pilot, and that he is under my orders. I
have not, of course, seen the paragraph; and, if I had, should not have taken
any notice of it. Better not to have credit for doing what you have done
than to have credit for doing what you have not done. Still I think, as the
Transcontinental exploration is purely a railway one, the name of some one
known in the railway world should be connected with it. Saw Mr. Inspector
Ahearn during the day; his fever is better, and he will be able to accompany
us to-morrow afternoon. He seems to know the country nearly all the way
to the Gulf, and says there should be no difficulty in driving all the way.
He knows Mr. F. Hann, and speaks well of him. The barometer has fallen
considerably since yesterday, and 1102 ft. yesterday reads 1250 ft. this
evening.

SUNDAY, FEBRUARY 6.—Writing all the forenoon some particulars to
Collier for Lukin, *re* Grand Trunk Line, and telegraphed to Mr. Palmer in
reply to his *in re* Hann. At 2 p.m., the barometer at Camp 14 read 1220 ft.
Got the horses together, and made a pretty fair start about 2 p.m., putting in
the two chesnuts as leaders. For many miles the country is open, undulating
plains, with little or nothing in the shape of timber or scrub. We followed
on the south bank of the Barcoo; at least we kept within sight of its timber.
There is not a blade of grass for miles, although the soil appears good; the
long drought seems to have killed everything. There is plenty of limestone
cropping up at intervals all the way. It was our intention to have done
to-day what we intended to have done yesterday, viz., to have gone twelve or
fifteen miles on the road to Blackall, where there is said to be good feed and
water. We went ahead in the waggon, looking back occasionally and viewing
the party. Presently the chesnuts began to show symptoms of distress, and soon
after gave in; so we took them out, intending to replace them; but, on looking
back, the pack-horses were nowhere to be seen. We waited and waited, and
at last saw them about two miles distant, parallel with us, but clearly on
another track. We tried to signal them by firing shots and waving handker-
chiefs, but without effect. They went on their way, no doubt thinking
they were pursuing us, as there were fresh tracks on the road. There was
then nothing for it but to yoke up the leaders again, and strike across the
country until we hit their track, and then follow it. This was done with great
difficulty, but shortly after our leaders were so completely done up that they
could not keep out of the way of the wheelers. We had, therefore, to cast
them adrift, and proceed with the two wheelers. We got on better thus for a
while, and sighted the party about a couple of miles ahead, evidently still
following what they thought our track as fast as they could; and now our
wheelers came to a dead stand-still. The driver then got on the best of them,

and went away to try and overtake the party, and bring back four fresh horses. It was 5.10 p.m. At 5.35 p.m. some horses came in sight, about a mile off, so there has not been much time lost, and the party cannot be very far ahead. The cook came with the driver, and took our remaining wheeler to ride after the two chesnuts, and we got clear under way at 5.50 p.m.; not such a very serious delay after all. We, in the waggon, reached the Camp, on Greendale Creek, just before it joins the Barcoo, at 6.10 p.m., and the barometer now reads 1225 ft. Fair grass and water; and yet, a week ago, before the rain came, I am told by the inspector of police, Ahearn, that it was just as bare as the ground near Tambo. The cook, with the two chesnuts, came into camp before 7 p.m., all tired, the cook carrying his saddle. This is Camp 15.

MONDAY, FEBRUARY 7.—Got a good start at 7.30 a.m., altogether. Barometer at 7 a.m. read 1115 ft. A beautiful morning after a nice cool night. The country, for the first two or three miles, is open plain, with a fair sprinkling of tolerable feed. The soil is good, but the drought has been severe. Then we enter some myall scrub, very lightly timbered—no timber worth anything. Very little grass, but looking as if, with very little more encouragement, it would burst forth with great luxuriance. There are even now some very pretty little patches of green. Two or three miles of this, and then alternating myall scrubby trees and open plains. There are not so many indications of stone for the first twelve miles as we have seen on other days ; and there is not a stick of timber suitable for railway purposes all the way ; even the trees on the banks of the Barcoo, which we have followed the whole distance, are stunted myall, &c., of little or no value. Now and then there are patches of water-worn, reddish sort of gravel ; I do not know the depth. The absence of timber and ballast is conspicuous. In other respects, a line may be made on the surface, with easy gradients, wherever you like ; and I may be mistaken about the ballast. We have halted a little in the shade, at 9.30 a.m., to let the others come up. The country looks a little more undulating ahead ; and on the spurs and knobs there may be some stone. After waiting half an hour we just sighted our party, and proceeded for about four or five miles over undulating, open plains, on the rises of which there was a show of limestone and a lot of water-worn pebbly stone spread over the surface. We came to some lightly-timbered country, and in a very short time we were on the south bank of the Barcoo. Here, as is the case of all other rivers, there are indications of rapid running floods. The bed of the river is mostly dry sand, with occasional waterholes. There are no evidences of very high floods, not more than fifteen or twenty feet above the bed of the river; but this would flood the country on the south side to a very considerable extent; the evidences are rather of rapid than very high floods. There are some large whitegums here on the river, and a few sleepers may be got from the myall and other trees if, indeed, the timber be of any durability. There is blackbutt and coolibar, but neither of them is of much use except for firewood. There is an hotel at Northampton, thirty-seven miles from Tambo, and thirty-eight miles from Blackall. We found a fair camping place on south bank of the river, about two miles west from Northampton, at about 1.30 p.m., and the barometer was 1120 ft. The police and Mr. Wyatt and Mr. Baynes remained behind to get some horses from Enniskillen Station. They arrived here about 7 p.m., having left the fresh horses in a paddock at the hotel, as they were troublesome. It may be that in estimating the rapidity of the water in the Barcoo and other rivers, I may not have made sufficient allowance for the extreme friability of the soil, which is much more disastrously acted upon by running water than even in Victoria.

TUESDAY, FEBRUARY 8.—At 7 a.m., the barometer at Camp 16 read 1020 ft. Started at 7.30 a.m. sharp, altogether, except the new horses, which had not come from the hotel. Nice open plains, and these lightly timbered, with patches of scrub. As we near Northampton Station, the country looks better; and, although there is not a great abundance of grass at present, there are evidences of plenty at times; there are, also, nice refreshing pools of water all the way for miles, but these are, I think, simply due to the late rains. We stopped about an hour at the station, and there, or near there, the new eight horses joined us. Rather a nice-looking lot. We also got one from the station, apparently a good one, but with the reputation of being able to buck. We crossed the bed of the river immediately after leaving the station, and proceeded on the north side. The land looks good enough, but the large open, level plains are completely bare: not a blade of grass, and very few trees or shrubs of any sort. Wretched-looking country at the present time; but, no doubt, well grassed sometimes. There is the appearance of stone here and there. I am still puzzled as to whether it is basalt or limestone. Some of it is coarse sandstone. No matter, it is good ballast. Waited an hour, about noon, on the road, under a shady tree, to try and catch sight of party, but could not. Came on to Chinaman's garden, four miles from Blackall. No sign of feed, but some on the river two miles off. The country continues flat and bare all the way, and there are occasionally appearances of stone; but the timber is very poor. Reached the Chinaman's garden at about 3.30 p.m. There is a nephew of J. Thornloe Smith's at Murphy's station, as storekeeper or something, called Dalby. At 4.15 p.m., the party turned up, but no inspector, nor Baynes, nor Wyatt. We fixed on a camping place, but the horses have to go two miles towards Blackall to the police-paddock. We cannot camp there because we cannot get near the water with the waggon; not nearer than we are here at this Camp 17. Barometer at 4.30 p.m., 1010 ft.; near Chinaman's garden, four miles from Blackall, and about a mile or so north of the River Barcoo, on a lagoon of dirty-looking water. We are camping under the advice of the constable in the absence of the inspector.

WEDNESDAY, FEBRUARY 9.—Two of the horses missing this morning, the two that came home late last night. Inspector Ahearn came to the camp about 10.30 a.m., bringing a lot of letters, some nearly a month old. He drove me to the township. Called at telegraph office and sent telegrams. Had lunch at O'Malley's Hotel, and remained to write some letters. Then drove home in Ahearn's buggy, which he has kindly lent me. A pretty heavy thunderstorm about 10 p.m.

THURSDAY, FEBRUARY 10.—In camp all day; writing letters and preparing second progress report. Mr. Wyatt and Mr. Baynes in Blackall.

FRIDAY, FEBRUARY 11.—As yesterday, trouble about horses constantly straying for want of feed. In Blackall in afternoon. Mr. Hann arrived by coach, and will come to camp to-morrow morning. Called at survey office to examine maps; telegraphed for some. Several letters (private).

SATURDAY, FEBRUARY 12.—Drove to Blackall in the morning. Mr. Hann came out to lunch and began to look into things. He fancies that, as he has another black boy, we can do without a groom, having assistance from Mr. Baynes and Mr. Wyatt. Drove to Blackall again in the afternoon. Saw Inspector Murray, arranged to shift camp to-morrow, for a short distance, to where there is better feed. Bad accounts about water and feed further north;

but travellers exaggerate. Mr. Hann brought me two thermometers, a prismatic compass, and a book of tables from Brisbane.

SUNDAY, FEBRUARY 13.—Writing letters for an hour or two in the morning. Mr. Hann came about 11.30 a.m. and had a look at some of the horses. At 1 p.m. we started the camp and drove to within about fifteen miles from Blackall, towards Aramac. I left the camp (No. 18) there, and went on about four miles to Inspector Murray's quarters. The country through which we have passed to-day is not very prepossessing; it is bare, but the soil is good, and there are remnants of "Mitchell grass," which only wants rain to bring forth feed in abundance. We saw very little stone on our way, but there must be some, as the formation is similar to that on which we saw lots of limestone and sandstone before. The timber is of very little value—in fact, all the way the want of timber is the great difficulty.

MONDAY, FEBRUARY 14.—Saw some limestone in Mr. Inspector Murray's garden, and found the place on the bank whence it came; there appears to be plenty there. Left Mr. Murray's at 9 a.m. Drove to Camp 18, near Junction Hotel, changed horses, and then drove with Mr. Hann into Blackall. Got telegrams from Mr. Morrisett about horses (thirteen) sent away from Roma to catch us at, say, Aramac, and from Mr. Grey about weather in the North. Called at different stores and paid bills, &c. Drove to camp in the evening; caught in a dust-storm on the way, and had a little rain.

TUESDAY, FEBRUARY 15.—Barometer, 680 ft.; thermometer, 85°, at 6 a.m. The horses had strayed and we did not get away till 8.30 a.m. All the pack-horses are fresh ones; and one, a determined buckjumper, struggled with his pack soon after we started, and eventually got rid of it. Mr. Hann returned Mr. Ahearn's buggy, leaving it at the Junction Hotel, to be called for to-morrow. For the first two or three miles it is open plain, and tolerable feed. Mitchell grass waiting for a little rain to be very plentiful; better now than in a good many places where we have camped. Then a mile or two of stunted scrub and no grass. About five or six miles from the camp we left this morning we crossed a considerable creek; not much water in it now. Can it be the Dismal Creek? Shortly afterwards another small one; then country looking better at every step, but little or no useful timber except at the creeks, where, perhaps, some sleepers might be obtained. There is some blackbutt and coolibar. At about seven miles there is a very fine waterhole, in a sort of creek; here we saw some sheep—the first stock we have seen to-day. For the first seven miles I did not see any stone, but afterwards, at frequent intervals along the road, limestone is cropping up, and I am sure plenty will be found. There is abundance of feed, and a good many sheep scattered about in small lots. Magnificent rolling-downs, extending for miles and miles, all equal to anything to be found in Riverina, or, indeed, anywhere else. Came upon some water about noon and made a halt, as we have fifteen miles further to go before camping. Altogether we have seen comparatively little stock considering it is such splendid country. After a spell of a couple of hours we started again. The country for the next twelve miles is not so good, at least the feed is not so. Stone still frequently cropping up, limestone, but some of it very soft and flakey; it peels off in cakes about one inch in thickness, like I have seen granite and basalt boulders do in Victoria, like an onion. At about 4.30 p.m. we found a pretty fair camping place, on what is called the Home Creek. Feed not very first-class, still a good deal better than some we have to be contented with. There is some very soft limestone on this creek; it will crumble with your fingers, but fizzes on the application of acid. Barometer, on reaching Camp 19, 700 ft.; thermometer, nearly all day in a box, about 87°; but a delightfully cool breeze. Our course has been nearly due north, and the distance about twenty-two or twenty-three miles. Found we had camped very near the Home Creek head station, belonging to Mr. George Fairbairn, of Melbourne, managed by a Mr. Johnston, assisted by Mr. Carey. From the map it is 145° 15′ E., and 23° 54′ S. It appears that what is called "triodia," on the map, is the same as the spinifex; and we are told we shall see lots of gidyah, large enough for sleepers, along the Alice River; it is said to be excellent timber for lasting and from 9 in. to 12 in. in diameter. Mr. Hann and I spent an hour at the station.

WEDNESDAY, FEBRUARY 16.—Barometer, 600 ft.; thermometer, 76¾°—at 7 a.m. Two of the horses missing; but I got away at 7.30 a.m., leaving the others to follow. For several miles the country is open, or very thinly-timbered plains, with plenty of limestone cropping up. The ground is bare near the track, but there are patches of feed not far off, and fine-looking downs in the distance. The soil is good, fit to grow anything; but what is to be done with the produce? Very little stock is to be seen anywhere, occasionally a few emus. We passed close to the head-station soon after leaving camp. It appears it is exclusively devoted to sheep, at least it is essentially a sheep-station. About nine miles from Home Creek we came upon a splendid waterhole on a creek that must be a tributary of the Alice River, and immediately we headed the waterhole we crossed the creek. The country is very bare, scarcely a blade of grass—too much travelled; but the soil is rich, as there is a fenced paddock close by where there is abundance of feed. We now pursue a north-east course, and I fancy parallel with the Alice River. I tried to find the river, but could not. A lot of saltbush in the timber, which fringes the plain over which our track runs. It appears that the large waterhole we saw at nine miles is on the Home Creek, which falls into the Alice River, but some distance off. After this, the country becomes a little more thickly timbered, and I fancy some of the timber may be useful. There is a good deal of a sort of box, and some very white-barked gum. I am afraid the latter is very soft. The country within five or six miles of our midday spell is very heavy for travelling—very sandy. It is called the desert; why, I do not know. There certainly is not much grass, but the trees and shrubs all look fresh and healthy enough. We found good water and good feed, and have only about eight miles to do after lunch. Started again at 1.45 p.m.; the country is slightly rangey for a mile or two, and no grass; then very sandy; but the trees look healthy, and we meet with the first spinifex or triodia grass—not bad looking. After a few miles we come to a small creek, in the waterworn parts of which are indications of gravel, say about 1′ 0″ thick, and also limestone cropping out. The country is more thickly timbered, and some of the timber might be useful; there is a sprinkling of ironbark, box, and bloodwood. To get here, we have made a considerable detour: the first day from Camp 18 we bore about 25° west of north; then to-day at least thirty miles considerably east of north. I cannot see why the line may not go straight from Blackall to Aramac. It is quite ten miles from where we had dinner to the Camp, No. 20. It has looked like rain all day. Hope to get to Aramac before it sets in. The Alice River, which we have followed for the last six or seven miles, evidently brings down a lot of water in time of flood. Our Camp No. 20 is on its south bank. Barometer on arrival, 680 ft.; thermometer, 85°. Barcaldine Station near here.

THURSDAY, FEBRUARY 17.—Barometer, 590 ft.; thermometer, 80°, at 7 a.m.

Got a pretty fair start at 7.30 a.m.; but one of the leading horses set to jibbing, and had to be changed into a wheeler. We crossed the River Alice immediately. Then, for a mile or so, the country is very bare and poor. Then we went through a patch of gidyah timber or scrub, about half a mile, miserable stuff, not a stick amongst it worth twopence. Then for miles over fine open and slightly undulating downs with plenty of feed, and a great lot of sheep spread about all the way. Occasional patches of scrubby timber and bushes, myall, gidyah and sandalwood. Not much stone showing on surface: occasional patches; and, from the nature of the soil, plenty not far off. We can see a long distance to the right, all the same sort of country; nothing, as far as I can judge, to prevent a line going straight from Blackall to Aramac. Met some drays, with a steam-engine and some machinery, about fifteen miles from camp, going to Barcaldine Station for sheep-washing. Came, about four or five miles further, to a dam; but no feed near it. It is pretty close to a station on which Mr. Brown is manager; but, as there is no feed near, decided upon going on twelve miles further after lunch. The plains or downs over which we have come to-day are as fine as any we have seen; but rain is very badly wanted. After the midday halt we came for about six miles over very fine plains, badly wanting rain. Then a little timber for a couple of miles, and poorer land. A couple of miles further, we crossed the telegraph line when it was bearing N. 20° E., our course being about N.E. We crossed it at a point twenty poles from where it had turned suddenly to the right in its course towards Aramac. Two miles further, and we came to a dam with plenty of water. Mr. Hann, on his way, heard that there had been a lot of rain north of Aramac, and that the Thompson was four miles wide. This was confirmed by a man we saw at the dam; but he says there is no feed near Aramac—none within twelve miles this side and seventeen miles the other. We must, therefore, go to twelve miles this side of Aramac to-morrow, and seventeen beyond on Saturday. The water in the dam is very slightly brackish. There is difficulty about getting meat. Barometer at Camp 21, 710 ft. Saw the first "flock" pigeons. A great lot came to the water, and nine were killed at a shot. I am told that sometimes thirty are killed at a shot. They are as tough as leather or the Gippsland lyrebird.

FRIDAY, FEBRUARY 18.—Camp No. 21. Barometer, 580 ft.; thermometer at 7 a.m., 85° This is Coreena Station, belonging to Mr. De Satgè. His boundary-rider came to breakfast, a character of fifty years' standing, once condemned to death for killing blacks, roasting piccaninnies alive, &c., and once five years for something else. Got a good start at 7.15 a.m., and came over fine open plains for about six miles. Have crossed the telegraph line twice, its bearing each time being about N. 10° E. There is not much feed, but the soil is good, only waiting for rain. There are indications of stone all the way, but no timber very near; and that which fringes the plain looks scrubby. We crossed a fence near this six miles from camp; and, shortly after, shot and skinned a young brown kangaroo. This occupied about half an hour, and by that time the packhorse party came up. For about a mile the open plains continue, then about a mile of miserable-looking scrub, myall, gidyah, &c.; but not a blade of grass. Then open plains for about four miles, when we reached the woolshed, which is said to be twelve miles from our last camp; and we have six miles further to go to our camping place, which is 12 miles from Aramac. The plain we have lately come over is not very well off for feed, evidently waiting for rain, as the soil appears good, and there is limestone. After passing the woolshed, the track makes a sharp turn to the left from about N.E. to N. 10° W. This continues for five

C

miles, when we turned sharp to the right for a mile, which brought us to a waterhole on a creek, said to be twelve miles from Aramac. This is Camp 22. Barometer at 11.30 a.m., 600 ft.; thermometer, 88° in box. The country from the woolshed is fine open plain; pretty fair feed in some places, but badly wanting rain. The camping-place is very bare, and the water dirty. Mr. Hann rode into Aramac, and returned about 9 p.m., bringing me a telegram from Mr. Barron, but no letters. We have to go fifteen miles beyond Aramac to-morrow for horse-feed. This Camp (22) is on the Aramac Creek. I find we can follow the Landsborough Creek, as there is feed and water.

SATURDAY, FEBRUARY 19.—Barometer, 540 ft.; thermometer, 78°, at 7.20 a.m. Got a good start at 7.40 a.m. Got out of the paddock at about 8 a.m. Then fine open plains several miles across, for about seven miles. Very little feed, but good soil waiting for rain. Then for several miles scrub, stunted myall, gidyah, and a few sandalwood bushes, but not a blade of grass to be seen. At about fourteen miles we came to Aramac, having crossed the line of telegraph twice at short intervals. At the first crossing it was bearing 320°, and at the second, 310°. Aramac is a decent little town; we got there about 11 a.m. Called at the bank and telegraph office, sent several telegrams, and arranged about the forwarding of future letters. Mr. Sheaffe, through some storekeeper at Aramac, wants to accompany the party; referred him to Colonial Secretary. Find we have fifteen miles further to go. Started about 2 p.m., after getting lunch at the hotel. The landlord lent me his buggy for a few days, and I drove it on to the camp, where, as usual, we found not a blade of grass, and none within a couple of miles. The police have not given us information as they ought. Got to Camp 23 about 5 p.m. Barometer, 680 ft.; thermometer, 99°. A long and tiresome journey. Find the coach starts for Withersfield at 5 a.m. on Wednesdays; and returns, starting from Withersfield, on Thursday. Engaged box-seat for next Wednesday. Stone on surface scarce all day, but it is below, no doubt. The country we came over, after lunch, for about fourteen miles, is good, open plains, little or no feed, suffering from drought; then miserable scrub, myall and gidyah—not a blade of grass, but good water. We crossed the Aramac Creek immediately before reaching the township; it has a lot of water at times, no doubt, like all the others. Our camp is close to the point where the Muttaburra and Bowen Downs roads diverge. There is a Chinaman's garden close at hand, from which vegetables may be got, cabbages, onions, radishes, &c. Anything can be made to grow anywhere; capital, babour, patience, and industry are all that is wanted.

SUNDAY, FEBRUARY 20.—Got up with the intention of doing a lot of writing to-day, but stupidly ate some bad sardines for breakfast, and in about an hour was taken dreadfully bad, vomiting and purging at a fearful rate. In about an hour I was completely prostrated, hardly able to stand without holding on. Not a drop of stimulant of any kind in the camp. Mr. Hann sent to the nearest out-station, "Stainburn," and fortunately got some whisky, which revived me; and I soon went to sleep, and by-and-bye awoke perfectly well, except being very weak. There was no writing done that day. Arranged that Mr. Hann should drive me to Aramac to-morrow, so as to be able to shift the camp on towards Muttaburra on Tuesday, as there is said to be better feed there.

MONDAY, FEBRUARY 21.—Left the camp at the junction of the Muttaburra and Bowen Downs roads, about fifteen miles from Aramac, having pretty well recovered from my illness of yesterday. Mr. Hann drove me to Aramac on my way to go by coach to Withersfield. This is a day sooner than I need have gone, but there is no feed where the camp now is, and it is desirable to get on to Mount Cornish. I occupied myself all the afternoon in writing up a lot of letters which were in arrears, in attending to telegrams, &c., &c. In the evening there was a little rain, and it looks as if there was to be a "fall," but not much came of it; there was, however, a good deal a few miles off. The barometer when I left the camp this morning read 490 feet.

TUESDAY, FEBRUARY 22.—Barometer this morning read 470 ft. In-doors all day, writing and telegraphing. Could not account for a telegram from Colonial Secretary asking me for report, until I ascertained that he had not received my No. 2, sent from Blackall. He could not have received it, because, as it was not posted until 11th, it could not leave Blackall before 16th, or reach Brisbane before 23rd or 24th. Received diary and letter-book from the camp in the evening, so as to be able to report from Blackall to Aramac if required, instead of from Blackall to Muttaburra. But can see now that there will be no necessity, as No. 2 Report will no doubt be received to-morrow.

WEDNESDAY, FEBRUARY 23.—Up before 5 a.m. The rain all gone off, but fancy there has been some towards the north. Got a good start by the coach at 6 a.m. Barometer, 510 ft. The first mile or so, as far as Pelican Creek, is open plain, good soil, but no grass. Then commences what is called the "Desert," a miserable country, open sandy plains, alternating with wretched gidyah, scrubby trees, and miserable sandalwood. Some of the gidyah trees are larger than any I have seen before, being ten or twelve inches in diameter, but very hollow and defective. I saw plenty before crossing the Pelican Creek, but none afterwards, and nothing but wretched sand; not a blade of grass anywhere; but there has been much more rain than at Aramac; lots of pools of water and lots of debris. They say that the country is at times flooded for miles to a depth three or four feet, and running fast. At fourteen miles we changed horses. No house, only a bush yard. Barometer, 550 ft. at 8.20 a.m. For about six miles after changing horses the country continues level, miserable desert, with a few salt-bush stumps here and there. At about six miles we pass out of the big paddock, and the country begins presently to rise. It is still very heavy travelling, sandy, but looking a little greener; the grass, however, is spinifex, said by some to be valueless, but by others to be very good feed when young, after being burned down by bush-fires, for horses and cattle. The timber is somewhat different. There are a few very white-barked gums and some good-sized gidyah trees. About eleven miles from end of first stage, or twenty-five miles from Aramac, we came upon some stone, a sort of sandstone, the first since Pelican Creek. The country all the way looks fresh and green, but still spinifex. We have been rising pretty regularly for the last six miles, and have now reached a height of 960 ft., then level to the end of the second stage, 28 miles from Aramac. The country here for the last two miles is very singular, with the appearance of having been a large lake for two or three miles in each direction. It is nearly level, and is surrounded, except at frequent breaks, by cliffs of sandstone, in some cases thirty or forty feet in perpendicular height. The stone varies; some is very soft, and there is mixed with it some conglomerate, also very soft. No doubt good ballast might be found amongst it, and without much trouble, because there is a good bold face. There is a garden here, in which potatoes and other vegetables grow readily.

The barometer at the refreshment place (we had dinner here at about 10.20 a.m., 2s. 6d.) at 11 a.m. read 960 ft. There is a government dam here. The horses which brought us over the first stage, fourteen miles, had to be driven here, fourteen miles more, to be fed on oats, &c., and those which brought us over the second stage had to be driven from here fourteen miles this morning. This is because there is no reliable water at the end of the first stage. By the time we got here the poor things were completely done up. There has scarcely been a stick of useful timber all the way from Aramac; still the ranges about here look as if they could supply some sleepers, &c. After starting again with a fresh team the ground rises 170 ft. in about two miles, making 1113 ft. It still continues to rise, but not so rapidly. At seven miles from Greyrock, where we changed, the elevation reached 1340 ft. This was the summit; it then falls rapidly fully 100 ft. in less than a mile. During the first two miles, the bluffs, whose tops are about horizontal, gradually died out. At the summit the same kind of sandstone formation shows up boldly. Up to this the land looks good, and capable of growing anything under cultivation, but there is spinifex grass all the way, and it really does not look bad. After passing the summit the descent at first is rapid, but then gradual for many miles, and there is a sprinkling of ironbark, from which a good many sleepers might be got. Then a few box-trees, and what is called the appletree, and now and then bloodwood. Grass is better; the spinifex seems to have disappeared. We arrived at the next place for changing horses at 1.45 p.m., but there were no horses there. Barometer at this place and time read 1000 ft. We could not wait for horses, and had to come on with the same team about five miles, to a hotel (Texas) at the "Dry Alice," where we took the horses out for half an hour and had another dinner (5s. 6d.) (We had had one at 10.30 at the Grey-rock.) We also fed the horses. We have six miles more to go to the next changing place. Level country, fair feed, and a good sprinkling of ironbark and box for several miles before we reached the "Dry Alice" and the Texas Hotel, kept by J. Richardson. Barometer at Texas at 3 p.m. equal 960 ft., same as at Greyrock. At changing place, between "Dry" and "Wet Alices," at 4.40 p.m., barometer, 1050 ft. We have come six miles through dead level country. Some ironbark, box and a sprinkling of curragong, with very good feed all the way. I have not seen any stone, but there is great abundance between the Greyrock and the Texas Hotel. From the last changing place to "Springers," the next stage, which is about ten miles, the country is flat. We ran up the Alice and presently crossed it, the river running from our right to our left, and all the way the country is gradually rising. There is plenty of timber and stone, and, although spinifex, the feed looks dry. The horses and cattle are fat, all from spinifex. The land is admirably adapted for cultivation, setting aside the question of market. We got to Springers, where there is a comfortable hotel, at 6 p.m, and decided to stop here instead of going twelve miles further on to-night, and to have an early start to make up in the morning. Barometer at 6 p.m. is 1230 ft. N.B.— A main line from Blackall to Aramac crosses the Alice river, the river running from right to left; a road from Aramac to Withersfield crosses the Jumpup Range, afterwards crossing the Alice River, running from left to right. It follows, therefore, that a line may be run from Withersfield to the main line without crossing either the Jumpup Range or Alice River. The waters of the Alice go into the Barcoo.

THURSDAY, FEBRUARY 24.—Barometer at 5.20 a.m., at starting, 1140 ft. Got a good start about 5.30 a.m.; and, with a ripping team, came over about ten miles to the Greenhill Hotel (kept by Mr. Penhallunck) by 6.45 a.m.

On our way we crossed the dividing range, at an elevation of about 1240 ft., but the rise is imperceptible, and so also is the fall.   At the "Greenhill" the barometer read 1040 ft.   There is plenty of ironbark timber and stone all the way, and the soil appears good.   There is abundance of feed, spinifex it may be; but the horses we have been driving are quite fat and frisky on spinifex.   Apart from this, it is good agricultural land, wherever circumstances require or justify it.   The water from here goes into the Burdekin River and into the sea at Townsville.   We had breakfast here at 7 p.m.   We passed a lot of green timber called "yellow jacket," said to be tough and good; it is of suitable size, 12 in. to 18 in. in diameter. From this to the next stage, "Rocky," we pass over rangey country, but the spurs are of no height.   For a considerable distance we follow the Rocky Creek.   The land is poor, but there is plenty of stone and timber, ironbark, yellow jacket, box, &c.; not much feed.   We reached Rocky at 8.40 a.m. (barometer, 905 ft.), twelve miles from last stage (at Greenhill Hotel).   At Sandy Creek at 10.30 a.m. (barometer, 830 ft.), over level ground.   Plenty of timber, bloodwood, ironbark, blackbutt; some stone; and fair feed and good soil.   The change of horses is one mile further on; a stockyard on the road.   Reached Surbiton (thirteen miles) at 1 p.m.   Barometer, 800 ft.   For the last seven or eight miles we have come over, which is called "Surbiton Flat," rich land and no timber, only a very few ironbark saplings.   There is basalt cropping up in many places, plenty of it, and Mount Surbiton, a short distance off, has no doubt plenty of basalt, at anyrate plenty of stone. Surbiton has only one house, the "Surbiton" Hotel; it is on the Companion Creek, before coming to the Surbiton Flats.   There was plenty of good timber, ironbark, hardwood, and blackbutt.   Arrived at the "Sunny Hills," where we stop, at 5.15 p.m., after a long, tiresome stage of about eighteen miles, through uninteresting and barren-looking country.   Plenty of stone, and, generally, plenty of timber, ironbark, bloodwood, blackbutt, gidyah, brigalow, sandalwood, &c.   On our way we crossed several water-courses, dry now, but evidently flooded sometimes.   We crossed the Companion Creek immediately after Surbiton; then, eight miles on, the Belyando River; and, in three miles more, the May Creek; both these are difficult.   Barometer, on arrival at 9.15 p.m., 930 ft. at Sunny Hills.   The stone on surface here is limestone.   We had taken off our boots and socks, and were preparing for bed, when, about 9 p.m., a message came to the effect that the coach from Withersfield towards Aramac had completely broken down forty miles from here, and the driver had ridden on to ask us to at once proceed to relieve the passengers.   There was nothing for it but to yoke up and that at once, which we did, travelling all through the dark night.   The moon did not get up before 2 a.m.   Of course, I saw nothing of the country till daylight, but it was all level and lightly-timbered.   We changed horses once or twice, but I was dreadfully tired and sleepy.   From daylight to 8 a.m. we travelled through nice-looking country, lightly timbered, with occasional patches of limestone cropping out.   We changed horses at the Teresa Creek, and I found, much to my surprise, we were crossing the Drummond Range.   You can scarcely call it a range at all; just a series of little spurs, which would not require seven ft. of bank or cutting.   The creeks no doubt must have attention; and there are several pretty considerable ones.   At about 8 a.m. we came upon the broken down coach and the passengers, who had been there for about eighteen hours.   We put them into our coach, and they started towards Aramac.   We have to wait whilst Johnny, "the driver," goes somewhere for a buggy to take us on.   I found there was an hotel six miles on, and decided to walk to it, so as to get some refreshment.   There were five passengers.   I and another walked on, the others remained.   The

highest point I noticed on the Drummond Range was 1040 ft., and Smith's
Hotel at Tomahawk Creek, which I reached at 10.45, read 950 ft. I got
some breakfast and had a good rest. The country is all ironbark, undulating
ground until within a mile of Smith's, then level plain with moderate brigalow
scrub. The soil all the way is good, and there is plenty of limestone and
sandstone conglomerate. Just exactly at 4 p.m. the buggy hove in sight, and
after the driver and passengers had had some refreshment, we made a start
with a capital team of horses. There are two routes; by one the distance is
twenty-four miles, by the other thirty-eight miles. We did the first stage,
about twelve miles, in very little more than an hour. About six miles on we
crossed a small creek, where there is plenty of granite cropping up. This is
the first granite I have seen. The country all the way looks splendid—fresh
and green, plenty of feed, moderately timbered, mostly with ironbark of sufficient size to be useful for railway purposes. At twelve miles stage a dis-
cussion arose as to whether we should take the new track, sixteen miles, or
the old one, twenty-eight miles. It was now 6 p.m., and as we could get
a guide we decided to try the new. I was afraid we would not reach
Withersfield by 8 p.m. We rushed along at a great pace for an hour and a
half, then it began to get dark. We crossed several large creeks fortunately
before it got dark. But presently things began to look queer; it was extremely difficult to find any track, and we had literally to grope our way.
We had fortunately amongst the passengers a Mr. Rowe, a squatter from the
Landsboro' Creek; but for him we must have camped, and after travelling as
we had been for the last twenty-four hours, under, in some cases, difficult
circumstances, we were not very fit. However, by-and-bye we struck the
main track, about what the driver called four miles from Withersfield. It
was a dreary long four miles, and we very nearly got upset more than once.
However, we got in just at 10 p.m., all pretty completely done up. It was
dark for the last two hours, and I could not see anything of the country, but
there appeared to be some very fine timber; we could occasionally feel some
stone too. Withersfield is a wretched place, or else I am very tired. The
barometer on our arrival read 725 ft. I got a pretty good sleep.

SATURDAY, FEBRUARY 26.—Barometer, 690 ft., say 700 ft. Got breakfast
in good time, and left Withersfield by train at about 8.30 a.m. They are not
particular to a few minutes. Found from the station-master that Mr.
Ballard, the chief engineer, had made provision for me to ride on the engine.
I see the sleepers are all half-round ironbark, seven feet long, and about
10 x 5 in. They are only two feet centre to centre. Here very little ballast
is used, only three or four inches under the sleepers. The ballast is of various
descriptions—basalt, freestone, limestone, and broken to 2½-in. gauge. The
rails are steel, and about 42 lbs. per yard. The gauge, 3 ft. 6 in. At
Emerald I was met by Mr. Ballard, and his principal assistant, Mr.
Hannan, who, I find, was at one time with my brother, G. H. W., on the
Kyneton deviation and other surveys. Mr. Ballard was exceedingly attentive; indeed, they both were. I had been afraid of a little jealousy. Nothing
of the sort. I came on the engine a great part of the way, and had an
opportunity of seeing some of the "low level" water provision. If I had any
hesitation before, I think there is none now. The Nogoa and Comet rivers
are crossed at the low level, the water passing over them to a depth of many
feet. I fancy Mr. Ballard said 16 to 20 feet. The gradients approaching
the bridges on either side are steep; I think 1 in 25 or 30; but they are
fairly well dealt with. Mr. Ballard has adopted the parabola for his change
of gradient as well as for his horizontal curves, instead of the circle, and with
great success. Riding on the engine I could not, of course, see much of the

bridges; but the impression left on my mind is that less timber would do the work, and offer less obstruction to the water. I think, too, the ironwork of the points would pay for examining. I do not think there is enough ironwork at the joints, or else there is too much timber all the way. Mr. Ballard intimated to me that he and Mr. Hannan had offered to undertake the survey of the Transcontinental Line. Had, in fact, told Mr. McIlwraith they were willing to do so. From what I saw of them both, I am sure they could do it successfully; it would be absurd to import English engineers for the survey, at anyrate; they would be nowhere. Mr. Ballard left me at "Dingo;" and, as we met what is called the "up"-train here, he returned to Emerald. After this I came on the engine with Mr. Hannan, and passed over a very remarkable flat. For about eight miles the line has been under water to a depth of 12 or 14 feet; not exactly running water, but back water. The great flood was in 1875. There were four or five lives lost, and Mr. Hannan pointed out to me where he buried the bodies, viz., at about the "flood level;" most absurd to look at it now. There are several dams on the line; I think, eight; some of masonry, others are simply the railway bank, with some precautions in the way of puddle-walls, &c. It is refreshing to see dams anywhere. Water is the great want of the colony. At Westwood I met the station-master, one of the first men who joined my survey party in 1854, Walter Thomson, a friend of Zeal's and Swyers'. Mr. Hannan left me when we reached his office on the line. We got to Rockhampton at about 6.30 p.m., and Mr. McEacharn was waiting for me on the platform. Several of the citizens had called and left cards; and I met the mayor and others during the evening, but the last two or three days of travelling have taken it out of me, and I was glad to retire early. The barometer read 150 ft. below zero, or 29.85 in. We had a smart shower on our way, about twenty miles from Rockhampton, and it looks like rain. I find from Mr. Ballard that the elevation of Withersfield is about 1000 feet above their datum, which is 100 feet below the doorstep of Rockhampton Gaol, or near "high-water level." My reading of Withersfield was about 700 feet, i.e., taking 927 feet for Charleville, and making no allowance for variation of the atmosphere. I find the sleepers cost from 3s. to 4s. 6d. each, and the ballast about the same per cubic yard. Mr. Ballard gave me a photograph map showing the country all the way from Blackall to Richmond Downs, on the Flinders River, except about 2° of latitude, i.e., between 20½° and 22½°; scale, 16 miles to one inch. From this I see a line has been surveyed from Withersfield to the main track from Blackall to Aramac, meeting it near Barcaldine Head Station, where we crossed the Alice River; and I am told it is a very easy section, excepting the crossing of the Drummond Range: the length, 367, − 197 miles = 170 miles. The telegraph line on this line from Blackall to Aramac is shown quite straight; and I can see no reason why the line should not be made quite straight. It would save a good many miles. There is a grand iron bridge over the Fitzroy River at Rockhampton just now completed. Mr. Byerly was the engineer. I must get some particulars from him.

SUNDAY, FEBRUARY 27.—Had a really good night's rest. Heard it raining several times, and it rained nearly all day. It had been arranged that I should be driven to Gracemere to have lunch with Mr. Archer and Mr. Byerly, &c.; but the weather was too bad. I spent most of the day with Mr. McEacharn and Mr. Lambert. To my surprise, found a card on my table, "H. Cadogan Campbell." It looks as if the drought had broken up, and the wet season set in. The barometer has fallen a little. I ought to be glad of the wet weather, but do not like the prospect of a coach journey in

the rain or its uncertainty; but there is no use in meeting difficulties half
way.

MONDAY, FEBRUARY 28.—Showers all day. A lot of people called upon
me. I was asked by the Chamber of Commerce to inspect the country from
Rockhampton to Deep Water, but declined, as my visit was unofficial; and,
besides, what I have seen of Mr. Ballard leads me to believe the matter would
be perfectly safe in his hands, ably assisted, as he would be, by Mr. Hannan.
My intended visit to the "lions" of the town—the hospital, the grammar-
school, orphanage, waterworks, &c., was postponed until to-morrow in consequence of the weather. I had lunch with Mr. Byerly, and a lot of the leading
citizens, at the Leichardt Hotel; and I had a walk over the Fitzroy Bridge
with Mr. Byerly. It is an imposing structure; but, of course, I could not go
into details. I believe it is the only suspension bridge in the colonies. I am
to have a description sent to me. Mr. McEacharn has given me a photograph
of it. The cost has been, I believe, about £53,000, and the length is nearly
1000 feet.

TUESDAY, MARCH 1.—Showery all day. Called at a photographer's, and
got some photos. taken. I drove round with the mayor, and a lot of the
leading citizens, to the various public institutions: the hospital, grammar-
school, orphanage, waterworks, &c. Cannot venture on a description. Everything most satisfactory, painfully clean and orderly. Then had luncheon
with his worship and some good company. Got some writing done in the
afternoon; and, in the evening, met a goodly array at a dinner given by Mr.
McEacharn. I cannot attempt to describe the hospitality of the Rockhampton people; it is simply "excessive," at the same time not "oppressive."
You may do as you like. I determined to get away by train to-morrow, notwithstanding telegrams to say the rain was heavy, and I would, probably,
be detained at Withersfield; but I want to be in a position to take the
earliest chance of getting forward. I wish to be with my party. I do not
like them to be camping out whilst I am under a roof.

WEDNESDAY, MARCH 2.—I got away by the first train in good time.
Several of my friends came to see me off. I am pleased to see them take an
interest in the work on which I am engaged; to this, perhaps, I have to
attribute their kindness to me. They evidently do feel an interest, and
they are instinctively kind and hospitable. I met Mr. Ballard at Emerald;
he is not at all well. I explained to him my views about the line from
Withersfield to the Main Line; he gave me satisfactory explanations as to
the choosing of the route and accepted my suggestions most kindly; he gave
me some further useful information, and some sections, &c. We got to
Withersfield in due course, and there is every appearance of some rain; perhaps it is unfortunate; I don't know.

THURSDAY, MARCH 3.—The weather looks dreadful; rain, rain, rain. It
cannot be helped, and perhaps it will help me on by-and-bye. But, oh! such
a wretched place to stop, and yet it might be worse; the roof is watertight,
the food is good, the beds are clean, the bedrooms are small, the way to which
is across a very muddy yard, and the rain patters provokingly on the iron
roof. Queensland is great for teaching patience. I don't think I have ever
been in such a fix before, and it has come about in such an unexpected and un-
accountable way. I never dreamt of it. In reply to a telegram sent to Aramac
this morning, I received :—"Raining since Saturday last, 6½ inches fallen.
No mails in from west. Rockhampton mails stopped here, unable to cross the

creeks. Believe rain general." This from the telegraph master at Aramac. There is some satisfaction in feeling that I could not move forward if I were with my party. The elevation of Withersfield, according to Mr. Ballard's section, is about 900 feet above high-water mark at Rockhampton, and my reading of the barometer at Jumpup Range is 610 feet above this, or say 1540 feet at summit.

FRIDAY, MARCH 4.—There appears no immediate prospect of the weather clearing, or of my getting away from this wretched hole ; perhaps it may be for a week, unless some special coach be sent, which is not at all likely. It was not my intention to have sent my report about the connection between the main line and Withersfield without mature consideration, and this is not the place for that. Among other nuisances is a horrid butcher's shop next door, from which the scent is intolerable. However, I set to work as well as I could and got the draft prepared. It is hard lines to be stuck up like this.

SATURDAY, MARCH 5.—Got my Progress Report No. 3, " Branch Line from Aramac to Withersfield," prepared. Do not feel at all satisfied with it. It has been prepared under very adverse circumstances. I passed the remainder of the day as best I could.

SUNDAY, MARCH 6.—There is no church, no nothing. The butcher's shop is partially closed, but they are giving it its weekly cleaning out, which is a great deal worse. They throw the brine out on the road in front, and the smell is almost worse than putrid meat. It all seems to be preparing me for some serious illness. The people of the inn are kind, but I am ill-tempered and out of sorts—impatient. There is no telegraphic communication to-day, and of course that is right.

MONDAY, MARCH 7.—There is to be no coach until Thursday, that is certain. The roads continue impassible to the northwards ; there is comfort in this, as I could do nothing if with my party, and it must give plenty of grass and water, and help our progress by-and-bye. I scarcely know how I passed the day. I went to Emerald by train, and found Mr. Ballard still suffering from bronchitis very badly.

TUESDAY, MARCH 8.—I must be going to have fever or something. I can do or think of nothing. Mr. Ballard devoted as much attention to me as his duties and his throat would allow. I got a good deal of information from Mr. Ballard about matters in general, but am greatly out of sorts.

WEDNESDAY, MARCH 9.—On my bed greater portion of the day, and examining some plans and sections. Came back to Withersfield, so as to be ready to start by the coach to-morrow. My complaint has culminated in what is locally known as the Belyando Spew. The " Belyando" is a river a little more than 200 miles west from Rockhampton, and its name will be remembered by me in connection with this sickness to my dying day. I never experienced anything like it. I vomited as long as there was anything to come up, and after that there was incessant retching. I could not keep down a drop or morsel of anything ; however, I had made up my mind to start by the coach, even if I perished by the way.

THURSDAY TO SATURDAY, MARCH 10, 11, 12.—I must lump three days together. I started by the coach inside, fortunately able to lie down on the seats, but most uncomfortably. The roads were in an awful state. The coach, I

D

dare say, was good enough, and suited to its purpose. The horses also good. The driver—well, he talked to the horses in a way that must have endangered his immortal soul, if he has any. This went on all day, from daylight till dark, day by day, for three days. We stopped at the most wretched inns at night. I could not look at food. I did sleep a little, not being murdered by the jolting of the coach. The journey came to an end, and we reached Aramac about 6 p.m., I think it would be an excusable exaggeration to say, "more dead than alive." Fortunately, I found a doctor here, who sent me to sleep. This ended the most miserable journey I ever had in my life.

SUNDAY, MARCH 13.—I was to have started by the coach this morning for Muttaburra. Under any circumstances, this would have been out of the question. I got a great lot of letters, some very old ones, none of any importance. Too many to read whilst I am so weak. However, the sleep is refreshing me, but I am not able to take any food. I continue in bed all day, and the Doctor (Sparks) is very attentive. I was fortunate in getting here.

MONDAY, MARCH 14.—A little better, and able to take a little beef-tea, &c. Before I left Aramac for Withersfield, on February 22nd, it was arranged that Mr. Hann was to meet me on Sunday, 6th March, about half way between Aramac and Muttaburra, and drive on in a buggy. Of course the floods told him I could not possibly come then ; but as he thought I might possibly come on the 13th, when he found I did not, he very prudently came on to Aramac, where he found me in bed. I cannot see that anything has been lost by my absence from the party. He remained in Aramac, and was of great assistance and comfort to me. I find this infernal " spew" is not at all uncommon, though it does not usually last so long. Mr. Hann seemed to know all about it.

TUESDAY, MARCH 15.—Able to sit up and do some writing to-day, and to make arrangements as to future route of party, management of horses, &c. Got my diary written, which has been neglected for some days. Was able to eat something. Arranged that Mr. Hann start for the party to-morrow at daylight, and I follow by coach on Sunday next, 20th ; he meeting me as arranged for the 6th inst. ; just a fortnight's delay.

WEDNESDAY, MARCH 16.—Mr. Hann started at daylight, and the arrangement we made yesterday is to be carried out as to my getting to Muttaburra. I continued to mend during the day, and was able to take some food. In the evening I got a telegram from the Honorable the Colonial Secretary, saying, " It does not kill." I say "it," for I hate to write the name. He expresses a hope that, now, with plenty of horses and good grass, I shall be able to go ahead soon. I hope so too. The horses from the first have not been suitable, and the grass and the water have been wretchedly bad. It is worth noticing that the water in this place—Aramac—is intensely hard, although it is rain water. It seems it is collected in a reservoir, the dam of which was made from a material taken from a pit properly sunk upon the upper side of it. There must be something in the material which affects the water. I think some analyses might be made, for the water might be unsuitable for engines, and something might be done to alter it. I have not noticed anything wrong with water before, except filth and stink from remaining stagnant.

THURSDAY, MARCH 17.—St. Patrick's Day, but I do not think the people here are very loyal Irish, for everything has been very orderly, and I have been able to examine my notes, so as to be able to make some progress with the draft for my Progress Report No 4, comprising the country from Blackall to Aramac. It is difficult to find material for report. The country is all good and suitable for railway construction; stone plentiful, timber scarce, &c., all as before. I think my reports in future will be very short. I am getting very much stronger. Replied to telegram from Colonial Secretary that "'Beyando Spew' does not kill," and explained route I proposed to take, viz., up east bank of Landsborough, then along Dividing Range to Cloncurry, &c.

FRIDAY, MARCH 18.—Got some information from a Mr. W. C. Jones that the route *via* Winton and Kynuna was much better. Telegraphed to the Colonial Secretary to this effect, and before he received my telegram he had heard the same from Scheaffe and telegraphed to me. Our telegrams crossed, and then he sent one explaining; so the matter is settled. During the day I made an examination of the effects of the late flood in the Aramac Creek. It is astonishing. The water must have been two miles wide, and now everything is as quiet as ever, except that there is a nice sensible stream running in the creek, and yet not a quiet sensible stream, for there are a lot of irregular anabranches in which water in considerable quantities is also running.

SATURDAY, MARCH 19.—Getting better all day, and preparing for a start by coach to-morrow. Settled with doctor, &c. Got things packed and all ready. Posted Report No. 4.

SUNDAY, MARCH 20.—Started with three other passengers by the coach, which is awfully crowded and full of luggage; a dreadfully rough road ; but, oh! what a change since the rain. The country is like an enormous meadow. I came by coach as far as Sardine Creek, thirty-five miles, and there found a buggy waiting to take me to Mount Cornish. I think to have gone all the way to Muttaburra would have killed me. I reached Mount Cornish at 4.45 p.m. There really is nothing to write about. Magnificent country, open plains, except about seven miles of gidyah and boree scrub from twelve miles to nine miles from Aramac. The Mount Cornish run begins at twenty miles. There is plenty of stone all the way. At Sardine Creek there is a fine government dam, and afterwards several constructed by the proprietors of the station. I scarcely know how to speak of the land. I dare say other downs that I passed over during the drought now look as well as these. There is scarcely a stick of timber within sight. There are three principal considerable creeks to cross, one at Stainburn station, and two Sardine Creeks, and there are many inferior ones. The water provision will require all the more care because the watercourses are not defined, and there will certainly be mistakes made, as there were on the Echuca and Beechworth lines, *i c.*, putting culverts in the wrong places. I cannot attempt to describe the kindness of the people at Mount Cornish.

MONDAY, MARCH 21.—A most miserable day, enough to drive one mad. It has nearly driven me mad. We had made up our minds and every preparation for a start on Tuesday morning; but this rain puts a stop to it, and fixes us here for an indefinite period. Sent telegrams over to Aramac, explaining matters and to let off a growl. The worst of it is we cannot send another telegram for a week in the ordinary course of things, no matter what may happen, unless a special messenger be sent to Aramac.

TUESDAY, MARCH 22.—The weather is looking better, and there is a possibility of our being able to get away to-morrow; at any rate, I have packed up all my things, got the bills ready, &c. During the morning I got Mr. Baynes to assist me in putting my accounts square, and we did it satisfactorily, from the time I left Melbourne until the time I reached Mount Cornish. I also sorted my things, taking only what I may actually require, and packing up the surplus to go to Brisbane by some opportunity. All is in a black portmanteau. We and it may possibly get to Brisbane about the same time.

WEDNESDAY, MARCH 23.—After the usual bucking and jibbing, which has always to be gone through after a spell, we got well away at about 9.30 a.m. All the country from Mount Cornish Huts to Muttaburra appears to have been flooded. We crossed a lot of anabranches, and, by-and-bye, the Thompson River, which has a bridge over it. After some time, and having crossed more anabranches, we passed over the Landsborough, which also has a bridge over it. Both are evidently submerged many feet in time of flood. I noticed, at one or two of the anabranches, that fording-places have been made with stone, about two or three inches gauge, quite loose, and yet not a stone has been disturbed by the late floods, although the water must have been several feet deep over them. It seems the Mount Cornish Creek is called the Thompson after the Towerhill Creek has joined it, and the Thompson and the Landsborough joined about three miles below Muttaburra. Below the junction would, no doubt, be the place for a bridge; but very great provision must be made for the water wherever it be built. I know of nothing in Victoria approaching it, the Wodonga Flat being the nearest, and that would not require half so much water-way. After leaving Muttaburra, the country is stony for a mile or so; then heavy sand, light scrub all the way for two or three miles, sandalwood, gidyah, &c.; then more lightly-timbered, with a mixture of boree. This goes on alternately with fine open plains until the junction of the Greenhill and Kensington roads is reached, about six or seven miles from Muttaburra. On our way I noticed some anthills, but did not see any on the road from Aramac to Mount Cornish. Plenty of stone all the way; a sort of broken-up and conglomerate. We had our lunch at the junction of the two roads; and, immediately afterwards, the country began to improve. Up to this it has not been prepossessing. Now it becomes more open, and presently fine open grassy plains, a little sprinkling of timber, clusters of small trees or scrub. About twenty miles from Muttaburra, we camped on Bangall Creek, in a good-sized paddock. By this time it was 5 p.m., and beginning to rain a little, but not threatening much. When passing through Muttaburra, I sent a telegram to the Colonial Secretary, informing him of our start under what I considered favourable circumstances. Copy to be sent to Collier. When they will go to Aramac I don't know, as the mail will not leave before next Tuesday. On our way we crossed several small watercourses, but none of any consequence. We saw Mount Leichardt all the way on our right, and Mount Mitchell at a greater distance on our left.

THURSDAY, MARCH 24.—Came on to rain during the night, and rather heavy from about 3 a.m. until daylight, when it got a little better. We determined to make an effort to push forward; indeed, it was no choice, for at this our present camp (No. 26), if a flood came, we must be washed away, as the ground on which we camped was so near the creek (Bangall) that it has all been under water during the late floods, although high and dry now. It is not pleasant preparing to start on a wet morning; however, we managed to get away about 8.30 a.m. As the roads were bound to be heavy, we put in the new team of strong horses. They are only intended for emergencies. They went splendidly; slow, but sure. We soon got out of the paddock, and

went on for miles of good country, fine open plains, on which sheep have been
feeding since the rain, and before the ground was dry. This detracts from
its appearance, but the soil is there. We came presently to some fine rich
flats, which have evidently been flooded recently. Then we crossed the
Bangall Creek again. The flats have been flooded to a considerable depth.
For a mile or two on each side of the creek, there are fences of wire across the
flat, and, at a considerable distance from the creek, there is debris left up
to the second or third wire; and, when there is the slightest hollow, imperceptible almost to the eye, the water has gone over the top rail and left its
mark. There is a dam near where we crossed the creek. It has been
damaged by the recent flood. The injury appears to me always to happen
at the ends, sufficient provision not being made for the overflow. It is very
difficult to make this provision, as, if a byewash be cut, the original surface
is disturbed, and the action of the water is frightful; the byewash would soon
be eaten down to the original level of the bed of the creek, in which case the
dam would merely cause a diversion of the creek. It might be done by
greater protection to the ends of the dam, by solid masonry or some way by
timber. This dam is built of stonework, with a puddle wall, to which I
think sufficient attention was not paid. There must be plenty of stone about;
there is plenty everywhere, I am sure; but there is that same scarcity
of timber, indeed, total absence for any useful purpose, except for firewood.
After crossing the creek, and getting over the ground subject to floods, we got
on to some magnificent downs, equal to any I have seen, with just a bunch of
scrub here and there. The rain was not heavy; indeed, it frequently stopped
altogether; but the road over the rich soil is very bad for travelling. It
came on to rain where we stopped for the midday halt, at about 1 p.m. It
continued to rain all the afternoon, and the ground became frightfully heavy.
Our brave horses still went on slowly, but surely. Without them we
could have done nothing to-day, so they have earned one-fifth their cost
already. We crossed Bradley's Creek about 2.30 p.m. We could not camp
there, as there were 16,000 travelling-sheep camped close at hand, so we
went some miles further on. We travelled at least eight miles more before we
found a camp capable of taking care of us for a week if another flood came,
which seemed not improbable. The country and the soil is all that could be
desired—rich rolling downs, as rich as any I have seen. There is a great
variety in the herbage; in some places it is good, and reminds one of a field
of wheat in May or June at home; in other places, sometimes for miles,
although the soil seems the same, the ground is choked with weeds, which
seem to have driven away and superseded the useful grasses altogether.
There is plenty of stone, but not a stick of useful timber, and nothing visible
in the distance for miles. The country seems to have been flooded for a very
long way on each side of every creek and watercourse; indeed, it looks to me
like a very dangerous country, more so than any other I have seen—a country
to be avoided if better can be found, and that a better can I doubt not.
I do not wish to be hasty in coming to a conclusion as to a general principle
which should guide the construction of the Transcontinental Railway, but I
am getting nearer to it every day. At the close of the day we had the
satisfaction of believing we were fifty miles from Muttaburra, or half-way to
Winton. When near the end of our day's journey, we met an unfortunate
carrier, with waggon and four-horse team camped. His horses were grazing
in the most lovely pasture; but the ground was subject to being flooded,
and he was just now, 5 p.m., thinking of yoking up so as to get on to some
safe ground. He had reason to dread the flood, for about three weeks during
the first flood, when attempting to cross the Western, within five miles of
Winton, the flood caught him, drowned five valuable horses, and destroyed

the whole of his freight, salt from the Aramac for Winton. The water went clean over the waggon, and, of course, the salt melted at once. We passed a good sprinkling of saltbush to-day.

FRIDAY, MARCH 25.—It has been raining pretty much during the night, but we had it dry for breakfast. We determined, if possible, to cross the East and West Darrs to-day, a distance of about fifteen miles. The roads are dreadfully heavy. We tried four of our best old horses, but they were no good, and had to fall back upon the new team, although they had all the hard work of yesterday. They, however, tackled it, hard as it was. All the way during the day we were travelling over rich brown soil, just wet enough to make the wheels clog. They first became like disc wheels, then the sticky stuff went on increasing until the wheels jammed against the sides of the waggon, and all came to a dead stop. The wheels then had to be cleaned with a shovel, a considerable work, and one that had to be performed many times during the day. It seemed at times hopeless to think of reaching the place for camping which we had fixed upon and set out for. The horses were fairly beaten, and seemed unable to stir, when a crack of the whip led the leaders to make a bound, which separated the connection between them and the pole, and away they went; their reins slipped through my hands like something greased. They did not, however, go far; they were easily caught, and seemed to see it was a mistake that had let them away. Damage was soon repaired, and fortunately the road soon became better, and we were able to proceed without stopping to breathe every half minute. We were now about seven miles from the camping place, and, until this change in the road came, the case appeared hopeless. We got on better now. At 4 p.m. we crossed the Darr River, one of those deceitful streams that it is impossible to provide for—a very insignificant stream, not at all indicating heavy floods, but looking as not to be trusted. We have passed over a lot of ground subject to floods to-day. Splendid soil, but bare of good feed; in some places there were good big bushes of good grass, but few and far between; in many other cases the ground was almost taken possession of by weeds; the soil, however, is fit for anything. I think this has been the hardest day, by far, that we have had, and yet we have only come fifteen miles; a mile an hour was quite the extent sometimes. Then we arrived at a miserable camping place, little or no wood, and what there is is too wet to burn. It has been raining all day, and it looks as if it meant to rain all night. There must be another flood coming, and this is such a miserable camp (No. 28).

SATURDAY, MARCH 26.—It did rain nearly all night, heavily at times; and appearances are decidedly against us. Mr. Hann is indefatigable, and, at about 8 a.m., started ahead to look for a better camping place; it would be of no use to shift from this except to go to a better one. This has been very miserable; but the delay for a few hours enables me to look over the maps and entries in my diary, &c., not altogether wasted. At about 11 a.m. Mr. Hann came back, having found a much better camping place about seven or eight miles west from this. It was at once decided to make for it. The country all around is frightfully boggy. The groom preferred to go seeking the horses on foot, without anything on except turned-up trousers. The poor little black-boy, "Pagan," is ill to-day, and will not say what is the matter with him; perhaps it is their way. I started in the waggon, at 12.30 p.m., leaving the rest to follow. We proceeded a couple of miles, the ground being slightly better than yesterday, then we got stuck; the ruts were so deep as to impede the waggon; the leaders broke their fastening again, and we had to stop for repairs for about half an hour. We then, after considerable plunging, "management," "persuasion," &c., got away, and, after crossing a rather deep creek, the first after the " imperceptible divide," we got on well for a couple of miles, thinking our difficulties were over for the day, and the party with Mr. Hann went on. But again we came to a heavy country, and the leaders got away, making a worse breakage than ever. More repairs, more patience required ; but no more repairs can be done. We are hopelessly bogged for the night, and can go no further without fresh harness; and then I am afraid the waggon will go. Mr. Hann came back, but could make nothing of it ; so the party is camped three miles on, and I decided to sleep in the waggon alone, sending forward such food as the men would require for the night, and keeping out enough for myself, for which very little will suffice. I retained a lantern to keep me company, but this brought all sorts of buzzing things into the waggon, and I preferred being quite alone. There was not a sound about ; there could not be, as there are no trees and no place where anything could rest, no shelter of any kind : a miserable, lonely, open, boggy plain, good soil but no grass. It is strange that these plains, for a great many miles, have very little useful vegetation on them, and yet the soil looks good, very similar to a great deal between Aramac and Mount Cornish, only more sandy, and subject to strange floodings. The water takes extraordinary freaks, going in all directions possible. I believe if a section were taken along the road, it would not at all show where the water in floods would cross it. There are no regular undulations which should guide the water. I fancy there must be a lot of shallow pans connecting with each other, and conveying the water in all sorts of possible and improbable directions ; and then there are lots of holes through which the water evidently gets away, and somewhat similar to crab-holes in Victoria. The country, so far, from Muttaburra is not desirable for a railway, if better could be found, and I know it can.

SUNDAY, MARCH 27.—I had been very much delighted during the night at seeing a good many stars about, and the weather altogether looking better. About 9 a.m. Mr. Wyatt came home, bringing some materials for kindling a fire to boil the kettle for the tea, which was on its way with some breakfast. Presently Joe, the driver, came in sight and I soon got something to eat. Mr. Wyatt remained with me all day ; he was occupied chiefly in copying my diary, &c., and I in writing letters. I find Mr. Hann started alone, leading a pack-horse, at daylight, to ride to the Vindex Station, to get something to pull us out. He thought the distance fifteen or twenty miles, though in the map it appears to be thirty miles. Joe came again in the evening with more food, and at sundown I was again left alone—so quiet—not a sound of a living thing except the twittering of some little bird or insect which I could not discover. There were some slight showers during the day, but the sunset was promising.

MONDAY, MARCH 28.—Bogged for forty hours, from 5 p.m. on Saturday to 9 a.m. on Monday, about half-way between Muttaburra and Winton. I had quite made up my mind to spend another day and night in this miserable hole, but about 8.30 a.m. horsemen and horses hove in sight. I had before this, about 6.30, got up, lighted my fire, boiled my tea, cooked my breakfast, and ate it, so I was prepared for anything. I found Mr. Hann rode seventy miles yesterday, i.e., thirty-five to and thirty-five from Vindex Station, to get something strong in the way of harness, &c.; determined to have me out of it. He brought chains and whipple-trees. The leaders were yoked singly, one before the other, and the four worked splendidly together ; we got clear away at 9 a.m., but the progress was terribly slow, the bogging and clogging of the wheels nearly as bad as ever. The wheels had to be cleared every

twenty or twenty-five yards; Tappington, the horse-minder, on one side,
Shekelton, the driver, on the other.   On we went in this tiresome fashion for
six mortal hours, and then we reached the camp; Mr. Hann with us all the
time, to assist in case of breakage, but there was very little.   The country
over which we passed was very much the same as for the last two or three
days : rich soil, plenty of stone, no timber, subject to very extraordinary floods.
I found we did not cross the divide until to-day, about a mile before we
reached the camp; like all other ranges, it is almost imperceptible.   This is a
much better camping place than the last two or three, but water will soon be
scarce.   There is no timber within sight—only a few bushes, a little larger than
gooseberry-bushes.   Barometer, on reaching Camp 29, when corrected with
Mr. Wyatt's, 780 ft., Charleville being 927 ft.

TUESDAY, MARCH 29.—Mr. Hann rode on a few miles first thing to
examine the road, and see whether or not there was any prospect of our
making any progress; it was decided there was not.   We remained in camp
all day ; the weather was fine for drying the ground, and we hope to get on
to-morrow.   I was engaged all day gathering materials for my Progress
Report No. 5, which must be written from Winton.   Got a tracing, a very
rough one, a skeleton from Muttaburra to Cloncurry, showing alternate lines
between Muttaburra and Kynuna.   Tiresome work waiting in camp when the
weather is fine; there is no help for it.   The rate of four miles in six hours
is not worth while.   The barometer fell a little during the early part of the
day, but recovered in the evening.   The sunset was good.

WEDNESDAY, MARCH 30.—Barometer, 28° 98', or 580 ft. against 780 ft. when
I got here on Monday ; this looks hopeful.   The horses had strayed and were
late coming in.   We got started about nine o'clock, with two of the good
horses as wheelers, and two of the old police greys as leaders, but they would
not do anything, and we had to make up the team with the four good draught
horses bought at Mount Cornish.   Then we got on splendidly, slowly of
course, but safely.   We had only to clean the wheels once before lunch,
which we stopped for at noon.   The ground is drying perceptibly every hour.
At about one mile from the camp we crossed the east boundary of the Vindex sheep station, belonging to Mr. Chirnside, of Werribee.   I think the east
boundary of the Vindex is the west boundary of the Darr Station.   Up to
the boundary the soil continues the same, rich brown or chocolate, but the
feed is still very poor.   Immediately after crossing the boundary (the fence
line is pegged out, but fence not erected), there is an immediate change.
The herbage is quite different, and goes on improving all the way.   First the
grass is as high as the horses' knees ; then, after a while, it is no exaggeration
to say, it is up to the horses' bellies.   We camped for the mid-day halt on
a tributary of the Western River, rather a branch of the Oondoroo Creek,
which falls into the Western at Vindex.   There is plenty of stone cropping
up at tolerably frequent intervals, but not a stick of useful timber within
sight, though the almost boundless plains are a perfect picture.   We crossed
a few tolerably well-defined water-courses, and the country is evidently far
less difficult to deal with than any of the previous forty miles.   From our
noon camp to the one for the night, No. 30, about five miles short of Vindex,
the country is all that any man could desire.   I don't mean to say the girth-
high grass continues all the way, but there is quite enough everywhere, and
the watercourses are well defined, and, though not large, are frequent.
There need be no difficulty in dealing with the water all the distance we
have come since lunch, about sixteen miles, making altogether twenty-four
miles to-day.   The ground is getting in capital condition.   The horses will

have a splendid night of it, as much feed as they can eat, and very little distance to go for it. It is only twenty miles to Winton, and by making an early start we may go as far we intend before camping for a mid-day meal. Barometer on reaching Camp 30, 29.18 in. or 400 ft. Fine sunset. Barometer continued to rise during the evening.

THURSDAY, MARCH 31.—Barometer, 29.30 or 310 ft. We got a comparative early start at 7.45 a.m., with a change of horses, as the roads are now pretty good. There was a lot of plunging and jibbing. No collars had been on for a month; however, after a little persuasion, we got away and slashed along nicely for about half a mile. Then came a piece of heavy ground, and the four gallant greys caved in. Then came more plunging, accompanied by breaking of harness; but we persevered, and were rewarded, for, after this was over, the "collars warmed," and they went as nicely as any team could go. About three miles on we came on five wool-drays, badly bogged, which had been there more than a week. The country so far is very beautiful, but the watercourses are not so well defined as they were yesterday. We went along the south side of the Oondoroo Creek, until we came opposite the Vindex Station, when we crossed it. We passed a lot of country which has been heavily flooded over the wire fences. If ever a line be made to Winton, it must keep on the north side of this creek. We called at station. It consists of ten ten-mile square-blocks = 1000 square miles, and is capable of carrying 250,000 sheep if fully stocked. It is splendid looking country. There is plenty of good feed all the way; not up to the horse's girths, but as much as might be consumed without trampling to waste a great deal. There is a very prominent range of hills extending for a length of thirty or forty miles. The foot of the nearest point is five miles from Vindex Station; and, as far as I can judge, its general bearing N.W. and S.E. It has no name, unless it be what is called the Opal Range on map. I sent Mr. Wyatt and Mr. Baynes to bring me a sample of the stone. After lunch we came along gaily for a couple of hours, then made our mid-day halt, quite certain of reaching Winton at 4 p.m.; but, when within a short distance of it, we heard, to our consternation, that Mills' Creek was flooded and impassable. Mr. Hann rode forward and returned to confirm the news, so we have to camp once more within a mile and a-half of Winton. Sent my letters into Winton, and two telegrams, one to the Colonial Secretary, the other to Collier. Oh! how tiresome it is. Barometer on reaching Camp 31, on Jessamine Creek, close to Mills' Creek, 325 ft., or 29.28. Mr. Wyatt and Baynes did not turn up at the camp; but I heard from Mr. Hann, who rode into town in the evening, that they had found their way to Winton. Fine sunset.

FRIDAY, APRIL 1.—Barometer, 260 ft., or 29.35. Mr. Hann rode over to the creek first thing to see whether or not the waters were subsiding, but soon returned to say that there was very little difference. Presently we saw a dray, which had been camped some days, going towards the creek, evidently intending to make an attempt. Mr. Hann went to watch them over. On his return he thought it not safe for me to attempt to cross in the waggon. I occupied myself for a couple of hours preparing draft of my Progress Report No. 5, Aramac to Winton. Handed it to Mr. Wyatt to be copied, he and Mr. Baynes having returned to the camp early. I then decided to go at once to Winton on horseback. Mr. Hann accompanied me. Our horses just managed to keep their feet on the ground all the way, not having to swim; but my legs were wet up to my knees. Mr. Hann returned to prepare for crossing the creek with the camp and party. I remained at Winton, writing letters, &c. About 4 p.m., Mr. Hann came to say all were safe over;

E

but he was very ill: an attack of fever, I think, producing retching and splitting headache, very much like what I had all the way from Witherstield to Aramac. Got some medicine from the chemist, but it does not drive away headache. I wish I knew something more about medicine. The mail from Muttaburra, which should have reached Winton on Wednesday, has not yet arrived, and cannot be heard of before it comes in the ordinary course of things: and, until it arrives, no mail can go out towards Brisbane. I met a Mr. De Kock, who is engaged sinking for water on stations. His terms are that, if he do not succeed in finding water, he makes no charge; but if he find a hole that will supply 10,000 gallons per diem, his charge is £250, and £50 more for a windmill and pumping apparatus for lifting that quantity. He estimates that this would supply 5000 sheep. I had no idea that each sheep required two gallons per diem. He tells me he used Wright and Edwards' machine first: but it was not strong enough, and he has now a much stronger one. The depth at which water is tapped is 90 feet to 100 feet; it is found in or underneath blue clay; there is no gravel. Thin seams of a sort of coal are sometimes met with, a sort of very dense lignite, but not exactly coal. Mr. Hann continued very bad all the evening. Barometer, 280 ft. or 29.31.

Saturday, April 2.—Barometer, 325 ft., or 29.27. Mr. Hann very much better this morning. Busy all the forenoon replenishing stores, getting harness repaired, &c. Decided to start after lunch, and go about seven miles, so as to have a fair start for Ayrshire Downs to-morrow morning. Telegraphed to Colonial Secretary. There is no mail in yet, and consequently none can go out. The harness took longer repairing than was anticipated, and we did not get away until 4.15 p.m.; we had an entirely new team, a little fresh to begin with. However, at about two miles we came to a very bad creek ; the horses would have taken us through, but one of the leaders slipped his fastening, and then it was all up. We tried them for a long time, and then had to fall back upon our emergency team. They took us out straight, and on to the camping place, about three miles further, to Bonnie Doon Creek, where there was feed, water, and firewood. We have now to look for all three. Until lately there was plenty of firewood everywhere ; now it is getting unpleasantly scarce. We reached the camp half an hour after sundown, having passed all the way over rich, undulating plains, full of feed and with occasional well-defined watercourses ; still evidences of flood extending a long way on each side of them. At the camp (33) there is a lot of stone, pebbly, on the surface, an unmistakeable flint. We are near the western foot of Mount Gordon. Barometer, 400 ft. or 29.20.

Sunday, April 3.—A lovely morning. Thermometer, 57'; barometer, 380 ft. or 29.22. All ready to start in excellent time, but we had a new team in the waggon, and, of course, there was the devil to pay. I thought we were never going to get away. One thing after another snapped until nearly all that was repaired yesterday wanted repairing again, but we got them away at last. We had not gone a mile when we came to a nasty creek. Our team decided not to do any more, especially the black leaders. We took them both out and put in a couple of greys—with no good result. Then we put in as leaders a couple of our emergencies, but the wheelers had sworn they would not pull, and they did not. So the emergency team had to go in, and in a few minutes the difficulty was over, and they were let loose again. We now had to repair damages. I think if Mr. Hann had had the blacksmith here who did the repairs at Winton, he would have crippled if not killed him. However, with rope, chains, and hobbles, we got started with our two grey leaders and the two original wheelers; then we did spank along, as the road was good, and we had to make up for lost time, exactly one hour. We got to Cookatoo Creek, about fifteen miles from our camp (33), at 1 p.m., and had lunch. Mr. De Kock was accompanying us. It seems the price, £50, for windmill and pumping apparatus is the price in Brisbane, and does not include carriage or fixing. About four miles from this we came to the first billabong of the Werna Creek. Of course the emergencies were called into use, and took us through this and the middle billabong, and the main creek, but it was nearly a case of swim. We then came on about nine miles, and on getting near the Wokingham Creek there were a lot of nasty billabongs to cross, so we put in the emergencies again. The Wokingham Creek has been my horror all day. I made up my mind to a regular soaking and nothing dry to put on; but we went through without the slightest difficulty. The water did not come within three inches of the bottom of the waggon. We camped close to the Ayrshire Downs Station, and congratulated ourselves on having done a good day's work, thirty miles, under rather adverse circumstances. But for the Mount Cornish team we never could have got on. The horses furnished by the police department are excellent in their way for fair weather and good roads, but utterly useless in cases of difficulty. For an expedition of this sort, good, staunch, strong, moderately heavy horses, able to travel twenty-five miles per diem, are requisite. During my trip from Muttaburra thus far, the question has often occurred to me—"What becomes of the sheep during the heavy rain, when these flat plains are either flooded or boggy?" The answer has come:—"900 wethers were bogged on Bowen Downs, and 5000 sheep were drowned near the junction of the Werna and Wokingham Creeks with the Diamantina River." The Cockatoo and Werna Creeks, although miles apart in drought, are joined in the floods, and dreadful havoc ensues; and, during the trip, I have come to the conclusion that the waters we have been crossing should all be crossed higher up, or headed altogether. Barometer, 140 ft. or 29.16.

MONDAY, APRIL 4.—Barometer, 390 ft. or 29.24. Considerable delay, repairing the damage done yesterday to whippletrees, &c. There is a smith's shop here, but no blacksmith. The horse-keeper, Tappington, however, is a bit of a smith, and made some capital strong connections. I had breakfast at the station, and sent a telegram to the Colonial Secretary reporting progress, but no one knows when it will go forward. We had our lunch soon after 11 a.m., as the smith's work had taken longer than we expected, and we want to avoid stopping again till the end of the day. For about four miles the country continues as it has been for a long way, rich rolling downs, but the herbage varies; then we get a little change for some miles abreast of the Lancewood Range. The surface is covered with pebbly, water-worn stones, and there is a great variety of saltbush, and some native caustic, which is said by some to be poisonous and injurious for sheep; but it heals sores and it burns slightly. After passing the range the quality of the soil improves, and soon becomes almost as good as ever. We crossed several nasty watercourses, very boggy, before we came to O'Brien's Creek, which nearly stumped us. We had not, however, to call upon our emergencies. Another team, greatly to our surprise, did it, and soon after we changed them: but to do this we had to go further into the police supply of horses, and there was the usual plunging and jibbing, legs over traces, furious kicking, etc.; but no hing broke, for a wonder. I think a good many of the weak points of the harness have been cured, and that, by the time we have replaced the whole of the original harness with repairs, we may have something reliable. At last we got away, and then really no teams could go better; for although we did

not get away until noon from Ayrshire Downs Station, we did twenty-five
miles and found a good camping place by 5.30 p.m.   There is, therefore,
every hope of our reaching Kynuna to-morrow afternoon.   Mr. De Koch has
accompanied us all day, and kindly shown us the track, without which we
should have experienced considerable difficulty.   Barometer, 470 ft. or 29.12.
Camp 35, on Watt's Creek.

TUESDAY, APRIL 5.—Thermometer, 61'; Barometer, 360 ft. or 29.24.
Got a good start on a fine morning at 7.40 a.m., with the four emergencies
in, as we had been told there was some very heavy ground for the first eight
or ten miles, until after the crossing of the Diamantina.   It was heavy travelling and no mistake ; heavy everywhere, but especially boggy at the
watercourses ; and for several chains on each side it was dreadful, but we got
through, and reached the first of the Diamantina billabongs about 10.30 a.m.
We crossed it and several more before we came to what I considered the
river proper; not that there was much difference between it and the others,
except that in it the water was running, and in the others not.   There are
flood-marks pretty high up in the scrubby trees, perhaps twenty or twenty-
five feet above the bed of the river ; and as the country all round is very flat,
the river must during the late flood have been three or four miles wide at
least.   Got to the old Dagworth Station, which is now deserted, about 11.30
a.m., and there we had our lunch and changed horses.   We met here a Mr.
Scafe, who lives on the "Great Dividing Range."   He gave us a not very
encouraging description of the difficulties we shall have to contend with
before we get to Cloncurry.   The country through which we have come to-
day consists of rich open plains, and plenty of grass, but most dangerous for
sheep in very wet weather, being very boggy.   There is stone cropping up
here and there, but there is no timber of any use ; even on the Diamantina
the trees are very scrubby.   Some posts might be got for fencing, and a
sleeper here and there, but not worth looking after, only fit for fire-wood.
About three miles after lunch we came to a very bad creek ; it is shown on
plan, but with no name.   It brings down a tremendous lot of water to the
Diamantina.   When the flood is up it must extend a mile each side, and be
twenty-five to thirty feet deep in centre.   Now it looks little or nothing.
It may be a billabong from the river ; there is debris high up in the scrubby
trees.   Here we got dead bogged and had to put in the emergencies, and they
took us through.   Then in about five miles we came to Crescent Creek, which
we had been told it would be impossible to cross, as there was a dray bogged
exactly in the middle of the track ; but we came through without any difficulty, and soon after released our "emergencies."   The flood marks are a
very long distance from the Crescent Creek.   Our new team brought us on
well until we came within quarter of a mile of where we were to camp, about
fourteen miles from Kynuna.   Here we got worse bogged than ever before,
because the horses were perfectly helpless, and all down, higgle-de-piggledy.
However, we got them free, and fell back again on the "emergencies."   In
the meantime we had placed a lot of branches, a good thick layer, for them to
start from, and then to walk on; and they, as usual, pulled us through, but not
before they had smashed some more of the original harness.   We got to a
nice camp at 6 p.m., where the sandflies and other flies were almost really enough
to drive one mad.   Mr. De Koek has been with us all day, and assisted us
materially in showing us the way.   We need not have crossed the Diamantina,
but there was no track on the eastern side.   The floods, at least the usual
remains of them, may be seen everywhere.   The body of water must have
been immense, both sides of the river being much alike.   The country through
which we have passed to-day is very level.   The soil is splendid, and any

quantity of feed; plenty of stone at intervals, but no timber. Camp 36, Barometer, 430 ft., or 29.15 at 6 p.m.

WEDNESDAY, APRIL 6.—Barometer, 308 ft., or 29.29; thermometer, 62°. Got a good start at 7.45 a.m.; and, to our surprise, the team went away without any difficulty; the leaders, however, were uneven. We came at a rattling pace for about half an hour; then we got dreadfully bogged, and, in the struggle, the pole snapped beyond all repair. This is about the worst fix we have been in. We got the "emergencies" to pull out the waggon; and, having found a sapling, a new pole was soon made, Tappenden's skill coming in handy. The repairs were finished by 11 a.m., when we dined. This mishap has sadly interfered with what we intended doing to-day; but misfortunes will happen, and do sometimes when least expected. We found some difficult watercourses to cross in the afternoon; but the police-horses did it well, and we got to the Kynuna Station about 5 p.m., a good fifteen miles if one. Here we camped for the night. Mr. Hann and I had dinner at the station, and learnt from Mr. Wilson that if we had tried to come up on the east side of the Diamantina, we should have been worse off than on the west side. He says we must go to Bell Kate to cross the river, and will probably find water about ten miles towards Cloncurry, after we have crossed the Dividing Range. It seems we have two or three boggy creeks to cross before reaching the river; but otherwise the road is good, and he fancies four days ought to take us to Cloncurry, with ordinary luck. The proper name of this station is Kynuna, after a large waterhole near. Mr. Wilson gave me a nice-looking piece of white sandstone, of which he says there is plenty. It is very soft when quarried, soft enough to be cut by an ordinary crosscut saw into slabs for paving, &c. Our camp (No. 37) is about half a mile from the station. Omitted to read the barometer. The country through which we have passed is very much the same as for several days past—rich soil, with plenty of feed; but terribly subject to floods. Thousands of square miles must have been covered with water during the late floods.

THURSDAY, APRIL 7.—Barometer, 305 ft., or 29.30; thermometer, 64°. We got a good start. Not quite so much plunging as usual, still some. Then we went away at a rattling pace for five or six miles, until we came to a creek that we were afraid to tackle without the "staunch team." All this occasions delay and annoyance. The weather looked threatening last night, but it is all right again to-day. The staunch team took us through without the least difficulty, and then we released them, and took the others back again. We had been told the distance from Kynuna to Bell Kate was seventeen miles. I am sure it is a good twenty miles. We reached the station at 12.40, having been travelling five hours. We find we can only go two miles beyond the river Diamantina after crossing to-day, as there is no water afterwards for seventeen miles, and the country very heavy. We shall then be beyond the Great Dividing Range. The country we have come over to-day is very flat; and, being pretty near the river, is subject to heavy flooding. The soil is as rich as any; but the feed, although beautiful, is not, I think, good; too many weeds with it. The stations about here seemed to be worked entirely by blacks. At Kynuna all are black, except the mailman and his wife; and the wife is a half-caste. There are a lot of gins and children, all nearly as naked as they can be. The men were away. Some other tribe had been encroaching, and the men had gone to fight them. The young gins are said to be very useful among horses and cattle; in fact, capital "stockmen." There was one little creature there at their camp only three weeks old. After dinner we got the strong team in, as there was plenty of difficult country and the river to cross. We at once commenced crossing a lot of nasty billabongs, deep and boggy; but all went well until we came to the last and worst, and then our pole snapped off like a carrot; fortunately, it was long enough still to answer, and, in about half an hour, we were under way. Mr. Hann, and all the packhorses, &c., had gone on to find a camping place. We followed their tracks as well and as long as we could ; but, by-and-bye, missed them ; fortunately, we were seen from the camp, and set right again before much time was lost. It then came out that the last and worst billabong that we crossed, when we broke our pole, was the Diamantina, and I had been looking for it for miles after we had left it behind. I think we have all had enough of the Diamantina. After crossing the river and its flats we came to rich, open plains, with nice dry spurs here and there. The soil is very rich, and there is stone in the mountains, of which there are several near here ; but there is no timber, except a little coolebar along the banks of the river. There is also sandstone all the way, apart from the mountain supply. I think the camp (No. 38) is about four or five miles north of the river. Barometer, 480 ft., or 39.10 at 5.30 p.m.

FRIDAY, APRIL 8.—Barometer, 420 ft. or 29.17; thermometer, 59°. Got a good start at 7.35 a.m. The first two or three miles pretty rough, not very; then we reached the Great Dividing Range, and the barometer read 410 ft. or 29.15. To call this the "Great" Dividing Range is a mistake, and calculated to mislead: it might be called the "Imperceptible Ridge Dividing the Great Waters," or the "Carpentaria Ridge." To the ordinary traveller, indeed to a careful observer, unless aided by instruments, it has scarcely any more material existence than the imaginary line which encircles the globe at its centre, or the dotted lines which separate the torrid from the temperate zones. After crossing it the ground begins to fall very slightly. For miles and miles nothing is seen but magnificent green pastures, with just sufficient patches of scrub to give it a park-like appearance ; there are also occasional small stony ridges, where the scrub is more plentiful and varied. There is sandstone cropping out, similar to that seen all the way from Roma. The soil is rich brown or chocolate, but the herbage is not good ; in many places it is completely choked with weeds. We crossed several rather nasty watercourses, and are in constant fear of coming to grief. Our pole is again showing symptoms of distress, and there is no beaten track. We are obliged to steer by sun and compass. We happened to come upon some water at 11.30 a.m., and then had dinner. Thermometer at noon, 71°. We were all ready for a start at 12.30 p.m. with a new team, and, of course, there was the usual kick-up and breakage ; the latter, however, was soon repaired—only the staple of the whippletree broken. We got clear away at 1 p.m., and, as the ground was fairly good, we made good progress, and must have done nearly five miles an hour for four hours. We camped on what we believe to be " Martin's Creek," near its junction with the M'Kinlay. The country has preserved the same park-like appearance all the way. The bushes are a little thicker in some places than in others, and the country in some places a little bit rangey, but very slight : stone all the way—sandstone, with lime in it. We had to cross two or three rather troublesome creeks and a great many watercourses, but did it without accident. There is not the same liability to floods as on the Diamentina. We found a good many new shrubs and flowers, some of them very pretty. The country I have come over since dinner will be common to all the three lines mentioned in my Report No. 5. It is all perfectly easy. There is very little earthwork, plenty of stone, but no suitable timber. Barometer, 405 ft. or 29.18 at 5 p.m., Camp 39. We crossed one point, this afternoon, where the barometer road 500 ft., the summit being only 440 ft. at the Divide.

SATURDAY, APRIL 9.—Thermometer, 61°; barometer, 300 ft. or 29.31. Got all started sharp at 8 a.m. Of course, with a fresh team there was a little difficulty, but not much. We were somewhat uncertain as to our exact position on Martin's Creek, but we struck out for N.W., and after nearly three hours' driving, during which we must have travelled twelve or fourteen miles, we reached Beau Desert Station (Hickson's), on the M'Kinlay Creek, five miles below the junction of Martin's Creek. Mr. Hann picked up Mr. Hickson on the way. All the morning we were travelling through the same rich, undulating plains, with occasional bushes, which look from a distance like fine large trees. Even in the creek (M'Kinlay's) there is nothing of any useful size. The country looks very beautiful everywhere, but the herbage is not, to my mind, very satisfactory. There is abundance of what I have called a weed taking possession of the ground everywhere. It is said to be eaten with avidity and profit by the horses and cattle when young. There is, besides this, a great variety of herbage, or weeds, or whatever it may be, and lots of creepers. Mr. Hickson proffered every assistance to enable us to cross the creek at the station, instead of going four or five miles down the creek and so much out of our way. By cutting away the bank a little, the strong team were able to pull us through. Then the country for about seven miles is one succession of billabongs. At times the creek is flooded, and, overflowing its banks, becomes about four miles wide, and fully thirty feet deep at its bed. The same magnificent country still continues. In places the soil is not quite so good ; the portion which has been flooded has caked over on the top, and the grass and weeds will, I fancy, soon show symptoms of drought ; but Mr. Hickson, the manager of the Beau Desert Station, told me the feed never failed there during the last dry time, though water became scarce. It is true they had some thunderstorms which did not extend to other parts of the colony. About seven miles from M'Kinlay Creek, as we considered the billabong difficulties over, we changed for a lighter team. The M'Kinlay Creek is much more defined than many of the rivers we have crossed. The best timber on it is, I think, coolebar, but too small to be of any use. Stone is plentiful : sandstone, with some lime in it. After changing our team, we came for about seven miles through thick, tall grass, preventing our travelling faster than a walk. Then came more frequent patches of gidyah scrub, and presently a lot of troublesome billabongs, which must belong to the Gidyah Creek ; very troublesome, because it took so long to find a place where we could cross. At 4.30 p.m. we came to a very bad one, and here, as we quite expected, our pole snapped ; did not exactly break off, but it was only by patching up with chains and very careful driving that we did about half a mile, when we came to a worse billabong than any seen before, and here, before crossing it, we camped for the night, and to put in a new pole before going any further. Barometer, 320 ft. or 29.27. This is Camp 40, on a branch of the Gidyah Creek.

SUNDAY, APRIL 10.—Barometer, 250 ft., or 29.25 ; thermometer, 62°. All the repairs, new pole, &c., were completed, and we got away at about 8.40 a.m. For hours we met with a succession of troublesome billabongs, taking up a great deal of time searching for crossing-places, and going in all sorts of directions. Fortunately our team—the strong one—never failed us. We started with gidyah scrub, which by-and-bye got thicker, and intermixed with other descriptions of scrub. Now and then we came upon a good-sized open plain, and on some of the creeks there is a little variety in the trees, if trees they may be called ; now and then a white gum, a bloodwood or a bau-hinea. Now and then we could go a mile without any interruption, but this did not often occur. The scrub gets thicker, difficult to drive through ; and there is little or no grass for a considerable distance. The soil, too, looks poor. We have made less progress per hour than for many days past. At about noon the country ahead looked better. We think we have done with Gidyah and Holy Joe's Creek, and after dinner hope soon to reach the Fullarton Creek, which is said to be a very nasty one; however, we have always surmounted our difficulties up to the present. After dinner we took an almost due north course, although, by going north-west we could have struck the Fullarton sooner. It appears that the road from Normanton to Cloncurry crosses the Fullarton about eighteen miles north from where we should have struck the river had we gone north-west. As it was, we struck it after travelling about twelve miles over very rough country, some middling scrub, some fine open plains with patches of scrubby timber, fair soil and plenty of feed; and then belts of very bad scrub, very difficult to get the waggon through; then plains again, lightly timbered. The country is very bad for travelling over, scarcely able anywhere to go faster than a walk. In fact, the day so far has been a most tedious one. About 4.15 p.m. we came upon the first of the Fullarton Billabongs, not by any means a formidable one, and an examination of the river itself showed it to be much better than expected; but we decided on putting in the "emergency team," and they took the waggon through billabong and creek without a moment's hesitation. An immense deal of water evidently comes down here in floods, extending to and connecting with the billabong, the distance between the two being quarter of a mile; but I do not think the country on either side is flooded to any great distance. Our course pretty much of the way from Beau Desert has been nearly N.N.W., and if we continue the same course to-morrow we ought to strike the road from Hughenden to Cloncurry in about six or seven miles, and we may probably, with great good luck, reach Cloncurry to-morrow night. We have seen very little stone to-day, and there is no timber good for anything except fencing. Barometer, at Camp 41, 400 ft. or 29.18.

MONDAY, APRIL 11.—Barometer, 320 ft., or 29.27, at 6 a.m. The horses —six of them—had strayed, and did not turn up at the camp until 8 a.m. As the waggon was all ready and horses in, we started at once, Mr. Hann leading, and leaving the pack-horses to follow. We took nearly a N.W. or N.N.W. course over fine open plains, occasionally dotted with clumps of bushes and small scrubby trees; there was good feed, but not so luxuriant as some we have seen; soil not quite so good. Then we came occasionally to belts of gidyah scrub, very difficult to get the waggon through; next, alternating with fine open plains. The feed on some of the plains is so thick and so high that at a little distance horses that have not packs on, and that are not ridden, are almost hidden from view. We had some very nasty creeks to cross, and could not make much progress on the plains beyond a walk, with, occasionally, a little spurt for a mile or so. After crossing a very nasty creek, which was cleverly done by four of the police horses, we stopped at noon for dinner. We had lost sight of the pack-horses for some time, but they put in an appearance by the time the tea was made. We consider we have come about fourteen miles this morning, keeping about a N.W. direction, in the hope of striking some track or the Williams River. We had great trouble in catching the waggon-horses after dinner, and Mr. Hann nearly came to serious grief over it. He had to lasso one of the wheelers, "Old Stumpy," and as he went forward towards him, after a great deal of pulling and struggling, the horse reared, made for him, and struck him full force with both fore-feet, one on the head, and the other on the shoulder. Down he went like a shot, and we were afraid he was killed; but it was only a case of "stunned." He soon got up, and, fastening the horse to a tree, punished him in such a way as only an angry man could. The horse struck at him again, in a most scientific way, with one of his fore-feet, but fortunately missed him, or it would have been serious. The horse clearly knew what he was about, and what he was punished for. At last he gave in, went quietly into harness, and behaved like a really good horse all the afternoon. In about five or six miles we came to Fullarton Creek, which was easily got over without changing horses. All the others, except Stumpy, were most troublesome at starting, but all had now quieted down. In about two or three miles more of magnificent country (the country has all been magnificent since dinner), we came to another creek (Elder's), which we knew nothing of before, and here it was deemed prudent to put in the emergency team, and right well they did it. It was by far the most difficult crossing we had met with. The banks were so steep, after getting down on one side, that we had to go along the bed of the river for about 200 yards, and then go up an incline as steep as the roof of a house. How we did it I don't know, but there wasn't a hitch. This I believe is the last stream of any consequence before we reach Cloncurry, which is said to be twenty miles distant; but no reliance can be placed upon "bush miles." We estimate we have come twenty-five miles to-day. Camp 42; barometer, 360 ft. or 29.24. This camp is on the west bank of the Elder River, not shown on plan.

TUESDAY, APRIL 12.—Barometer, 250 ft., or 29.35; thermometer, 65°, at 7 a.m. Got a good start at 7.45 a.m. A lovely morning, and very little plunging or jibbing. Country pretty level, open plains, with a sprinkling of bushes. Plenty of grass, mixed with spinifex; and, after five or six miles, spinifex seems to gain the ascendant. It is very tall, and does not bear very favorable comparison with the grasses I have been travelling through lately. We crossed several little watercourses. There are occasional ant-hills, and there have been a few—very few—all the way from McKinlay's Creek. At about nine or ten miles from the Elder, we came upon a very nasty creek. It took more than half an hour to find a tolerable crossing-place; but we got clear of it, and hoped it might be the end of this sort of thing; but about two miles on we came upon the worst creek of the lot, not by any means the largest, but so deep and so narrow that we should be quite certain to break the pole if we attempted to cross without letting down the waggon by hand; so we cut away the ground, removed the horses, locked the hind-wheels, and let the waggon skid down. The "emergencies" then took it out easily on the other side. We have been coming through the most "luxuriant spinifex" for a long time, and, soon after crossing the last creek, the ground began to rise and look very poor, covered with quartz and other water-worn pebbly stones. We came upon a track at last; and, on it, crossed a creek. We now stopped for dinner on some very dirty water. Mr. Hann thinks this must be Bishop's or Fisher's Creek, and we could not be more than a few miles from Cloncurry. In this we were very much mistaken. After a short time, we surmounted a ridge, from which Mr. Hann pointed out to me a mountain, at whose foot the copper mines were. I estimated it at fifteen miles distant, and a little west of south. We had evidently come much further north than we imagined. It was now four o'clock, but we determined to push on; indeed, there was no help for it, as there was not a drop of water with the party. Mr. Hann went ahead to search every gully for water, but without success. The sun went down, and the twilight was over, and still no water. Fortunately there was a good moon; and, by its light, we determined to strike out south-west for the road from Fort Constantine to Cloncurry, which follows the river. At about 8 p.m., we struck the road. Mr. Hann at once rode on to seek the township, and we followed at a smart

F

trot. By-and-bye we met him. He had been to the township, and now took us to the river to camp for the night. It was a most fortunate thing there was moonlight, or we should have been in a sad plight on the open plains all night, without a drink of any kind; but all ended well, and we were secure at 9 p.m. Mr. Hann found there were no letters at the post-office. No mail had come in for several weeks, but one was expected on Thursday next.

WEDNESDAY, APRIL 13.—Barometer, 320 ft., or 29.27. On inquiry and consideration, I find we did precisely the right thing in heading the range yesterday. It extends from the copper mine to the front, where we headed it a distance of about sixteen miles, and could not be crossed by a line direct from Beau Desert to Cloncurry. In fact, the main line cannot come nearer to Cloncurry than within fifteen miles; but the country is admirably adapted for a branch. I was busy for two or three hours in the morning preparing my Sixth Progress Report; then rode into the township; and, with Mr. Hann and others, went to the mine. Mr. Henry is one of the proprietors. It really is a wonderful sight. The richness of the ore is amazing, a great deal of it yielding 50 per cent. of copper; and some, but only small pieces, being pure copper. The quantity appears to be unlimited. (I am guided by Mr. Henry's representations.) A good deal has been sent to Sydney for smelting. It has fetched about £28 per ton, and cost—

Raising, say	£0	5	0
Land carriage to Normanton ...	8	10	0
Freight to Sydney	1	10	0
	£10	5	0
Worth there	28	0	0
Leaving profit ...	£17	15	0

A very nice little thing, and quite legitimate, if true. The mine is not being worked at present; but there is about £40,000 worth (Mr. Henry's state-ment) of ore on the grass. The cause of the stoppage was the failure of one of the London partners, a Mr. Colley. We, i.e., Mr. Wyatt, Mr. Baynes, and I, were engaged writing all the rest of the day at the hotel. We had intended to start in the afternoon; but, as we could not get the reports finished soon enough, it was decided to postpone the start until early to-morrow morning. All the neighbourhood seems filled with mineral wealth, and will, I think, be the principal feeder for the railway. It will, certainly, when developed, bring enormous revenue in the shape of freights. I cannot form any idea of how many tons per week will have to be carried. There are, also, besides copper, mountains of rich ironstone all around. I posted my Report, No. 6, and telegrams to the Colonial Secretary and Mr. Collier. Returned to camp at 5 p.m. On examining the waggon preparatory to starting to-morrow morning, we found some of the ironwork broken, which we had not noticed before, making things ugly. Barometer, 420 ft., or 29.16, at 6 p.m.

THURSDAY, APRIL 14.—Barometer, 290 ft. or 29.32. A further examina-tion of the waggon this morning decided us that it must go to the black-smith, and we must lose another day. It cannot be helped. Better lose this day here than a week by-and-bye, through a break-down far away from any repairs. There is no smith between this and Burketown. Mr. Henry called at camp to say good-bye, and seemed pleased at our not going. He gave me a lot of information about copper mines, with several of which he appears to be connected. I promised I would do all I could to help him if he attempted to float a company anywhere in Victoria. The development of this mining industry would help to settle the colony, and bring lots of freight to the railway. I tried for some time to do writing under a mosquito net, but the puffing wind upset everything. It appears there is a great variation in the yield of copper from different mines. Some will yield 80 per cent., or will be pure copper, when others will give only 12 to 15 per cent. Mr. Henry thinks that probably this mine will average about 30 per cent. I remained at the camp all day writing, examining plans, marking routes, &c. It appears to me that a straight line may be run from Aramac to the point where we headed the range, sixteen miles from Cloncurry, going very close to Muttaburra, missing the Diamentina, and going through good and easy country all the way. It may continue on straight to within a very short distance of crossing the Cloncurry River. A mail has come in from Hughenden, but brought no letters for us. No mail has come from Normanton, as the Flinders River is flooded ; and, as the mail has not come in, our letters and telegrams, which we posted yesterday, cannot go forward. Mr. Wyatt and Mr. Baynes during the day went to a mountain a few miles from the camp, and found it one mass of iron ore. They brought some very rich samples. Smelting, or fuel for it, is the difficulty both for iron and copper. I am told it would take four tons of coal to smelt one ton of copper. Now suppose the ore to yield 50 per cent. of copper, it would, it follows, be better to take two tons of ore to the coal at Sydney than to bring four tons of coal to the mine, and then have to carry one ton of copper to Sydney. The waggon came home about sundown, most effectually repaired, and fit for Point Parker now, I think. Got *Queenslander* of 12th and 19th of March in evening, through courtesy of police.

GOOD FRIDAY, APRIL 15.—Barometer, 205 ft., or 29.41 ; thermometer, 61½°. Horses a little bit troublesome after their rest yesterday, but we got clear away at 8 a.m., and, after more than the usual exuberance, we came on at a spanking pace for about sixteen miles, when we turned off the road to get a crossing of the river. On our way we met a Mr. Holt, who had been travelling from Normanton since 20th of February, and had been stopped by floods on an island for several days, with nothing or little, except wallaby or opossums, to eat. The country all the way is very flat, and splendid for travelling, a capital road ; and on each side are nice-looking plains. On the west, reaching to Cloncurry, and on the east for many miles, the country is looking very beautiful, with abundance of feed ; but the soil is variable, and not on the whole first-class. There is a good deal of spinifex, a few bushes here and there, and by-and-bye a little more timber, mostly gidyah. There are some white gums, bloodwood, bauhinea, beefwood, &c. A few sleepers might be got, but scarcely worth looking after. Stone plentiful and a little various, chiefly sandstone ; but there is what looks like limestone, nearly white, and unmistakable granite about twelve miles from Cloncurry. The river, which we soon reached, must be a sight in time of flood. There is very little water running in it now, only a small channel in the middle : but the bed of the river, which is about 300 yards wide, is all sand. The banks on each side are not very precipitous. The "emergencies" soon dealt with it easily, and at 12.40 we were over. Went straight at right angles to both sides of river. The bank on the west side is the highest, but I do not think this is more than thirty feet above the bed. I cannot see remains of any flood over this bank, but Mr. Hann informs me that in 1873 it was many feet above it, and the stock-yards which are near here and on high ground were washed away. This is the Fort Constantine Station. We passed a mile or so to the south of the rock after which it is named, because it was considered by some one who was in the Russian war to be like some fort he saw there. We have been travelling fast all the morning, but the pack-horses came on to us within half an hour, and then we had dinner under the delightful shade of a mature bauhinea tree, a most refreshing cool breeze having been blowing all the morning, the air being delicious. After lunch we put in the same team we had in the morning, but the ground was so rough that we soon had to make a change for some stronger ones. It is nasty country to travel over, terribly jolty, with lots of pits, like crab-holes. It is level, open plains all the way, fair soil, plenty of feed, but a mixture of spinifex and a good many scrubby bushes for the first four or five miles. Near the river are a good many white gumtrees, and some bloodwood. On the second four or five miles there is less scrub, and then we came to a creek which called for the use of the "emergencies." Our progress has been slow this afternoon, but we did first-rate in the morning. At about half-past four we came to a very nasty creek, and had to put in the "emergencies." Here we found a bloodwood tree (broad arrow over XCIV). It is one of Ward's survey marks. With a little excavation, we got through the creek and camped for the night, having done about twenty-seven or twenty-eight miles during the day. This is Camp 44. Barometer, 220 ft. or 29.38. The creek we have just crossed is Tommy's Creek.

SATURDAY, APRIL 16.—Barometer, 140 ft., or 29.47; thermometer, 62°. We got a good start at 7.40 a.m.; and, for a wonder, little or no plunging. We have been steering nearly due west ever since crossing the Cloncurry, having for a mark a prominent mountain after about three or four miles. This morning we passed it. Its formation is peculiar: a very hard stone, looking as if it had fire for its formation, still veiny; it is in small greyish brown and black layers; very brittle, but I should think very durable. There are many similar hills within sight, and any quantity of ballast for scores of miles might be procured here. After three or four miles more, we came to a rather formidable creek, on which we could not find a suitable crossing-place; so we followed it down to its junction with the Williams River, and found a crossing of the river below the junction. The Williams is not nearly half the size of the Cloncurry. It is strange that nearly all the rivers are more easy to cross than many of the creeks. This river seems remarkably free from billabongs. Our ordinary team took the waggon across without difficulty. Approaching the river, and, after crossing it, the timber is larger, especially on the western side. There are some bloodwood and whitegums that would cut at least a dozen sleepers each; but I do not think that the whitegum is of any use. After leaving the river, we soon came upon some good ground, and had a smart trot for a couple of miles; then, for three or four miles, the worst jolting we have had yet: one succession of beastly little pits, crabholes, &c.; we came over fine, large, level plains, with occasional belts and patches of scrubby timber. The soil seems pretty good but patchy, and the feed plentiful enough now, but coarse; in fact, I fancy all the country from McKinlay's Creek is better adapted for cattle than for sheep. There is stone occasionally on the surface. At noon we stopped for dinner on a little creek without any name. There is plenty of granite in a range to the north of the plain over which we have come, and some loose bits at the foot of the mountain which we passed. We had a new team after dinner; but the leaders did not like the uneven, jolting ground which we have had all the day, and after a few miles we had to change them. Still our progress was slow, scarcely

better than a walk; and, at that, nearly enough to shake one's life out. We
began to find, too, that, guided by Mr. Henry's information, we were out in
our distance, and began to fear we should not reach the Dugald river before
sundown. There seemed to be a prospect of camping out, and, of course,
there was not a drop of water with the party; but we made a spurt, and presently came upon unmistakable cattle tracks leading to water not far
off. They became more and more cheering, and presently Mr. Hann,
who had gone ahead, gave a signal by lighting a fire, and we were
soon relieved from our anxiety. The river Dugald is not a large one,
and we got across with our ordinary team just about sundown. It was a
wretched place for feed for the horses. We saw the first cabbagetree-palms
on this river. As we neared the river, the scrub and timber became a little
more dense; but, until then, we had the same open plains, with plenty of
grass, and a considerable mixture of spinifex. The soil is good, but dreadfully uneven, making the waggon creak, distressing the horses, and nearly
killing us. There is lots of granite, large boulders and hills, thirty or forty
feet high, composed of very large blocks. Some is hard and good, and some
rotten, soft, flakey, scarcely like granite at all. Barometer, 205 ft., or 29.41,
at 6 p.m. Camp 45.

EASTER DAY, APRIL 17.—Barometer, 190 ft., or 29.42. Got a fair start
at 8 a.m., and followed a nearly northerly course, although our proper course
would have been about north-west. The object was to strike a dray-track
running nearly east and west, nearly north from here about ten miles. The
first part of our journey was rangy, slight undulations, not at all objection-
able for a railway, but rather an advantage, as it can be seen clearly where
provision for water *must* be made. The surface is covered nearly all the way
with pebbly quartz and granite, and there is a little scrubby timber and some
bushes. At about ten miles we got out of the rangy country, and into a
tolerably level, lightly-timbered plain, with lots of appletree saplings, some
whitegum, and smooth barked gums, very brown, nearly red in color. Here
we found the dray-track, which was of great assistance to us. We can travel
three times as fast with a track in this country as we can without. After
about two or three miles we came to the commencement of more ranges, the
undulations very slight, and the ground falling westward towards the
Leichardt river. The barometer read 260 ft.; and, after two or three miles,
we were through the ranges, and it read 200 ft. We were now on to nice
undulating plains, lightly-timbered with gidyah, box, apple, bauhinea,
gum, &c., mostly saplings. We came along the track over the plains three
or four miles; and then the barometer read 160 ft. Here we came upon
water within six miles of the Leichardt, and had dinner; but it was nearly
an hour before the pack-horse party came up. As we were so near the river
we put in the strong team at once. We found some cabbagetree-palms last
night on the Dugald. I never tasted any before. I think it delicious; as
nice as any filbert I ever tasted, and the stem where it is cut is about three
inches in diameter. This is immediately below where the branches spring out;
and it is good for about eighteen inches or two feet downwards from this.
I think one might live for days on it. About a mile after dinner we came to
the Cabbagetree Creek, a very nasty one, very steep on both sides; however,
we had the "emergencies" in, and it was soon over, but immediately afterwards
the whippletree broke. It was fortunate this did not happen at the very
steep incline coming up from the creek. In about four or five miles we
reached the eastern bank of the Leichardt river, a most formidable affair. It
has evidently been a puzzler to many others, for there are the remains of huts
and other things that tell of long camping. There had been a crossing

descent excavated by previous travellers; but running water had got on to it, and it was in a very dilapidated condition; however, pick and shovel, and hearty good will soon repaired it, and we proceeded. We were clear across in forty minutes from the time we halted. The river is, I think, more than a quarter of a mile wide, sandy bed, steep incline on east side; better on west. There are a great many cabbagetree-palms, some fifty feet high, and a new tree which I do not recollect seeing before. It is an eucalyptus, very large. Each tree would turn out thirty to fifty sleepers, and the timber looks hard and good. There are also some good-sized whitegums; plenty of feed all the way, and pretty good-looking soil; but a good deal of spinifex. Stone plentiful—all granite, or nearly so. Some gravelly conglomerate, cemented firmly together. Found a good camp (No. 46) at 5 p.m. There are sparsely distributed anthills all the way from McKinlay Creek to the Leichardt River.

MONDAY, APRIL 18.—Barometer, 150 ft. or 29.47.   Two of the horses missing; this delayed us for a time, but we got away at 8.30 a.m., including the usual and indispensable kicking and plunging. The country is level, at least falling uniformly, and we followed the west bank or nearly of the Leichardt. Plenty of good feed all the way, and the soil seems pretty good. There is a peculiar grass here whose seeds are the terror of horsemen; and I am sure would be the death of sheep ; the seeds would eat into and go completely through them. The seed is about five feet from the ground, just convenient for catching a horseman's knees, and when once it gets its nose into his trousers or anything it works its way forward. We came to the place at which it is proposed to re-cross the Leichardt at 10.15 a.m.; so the distance cannot be more than ten miles, and we had expected fifteen miles. I have not noticed any stone this morning. We had to wait here for the "emergencies" and the pack-horses. To pass the time Mr. Hann cut down a cabbagetree-palm. I had no idea there was so much labour in getting the cabbage ; it took him fully half an hour with the little assistance I was able to give with a tomahawk. The tree was about twenty-five feet high and ten or twelve inches through at the bottom. After the tree was down, several of the lower branches had to be cut away before the real cabbage, fit to eat, was got at ; then, when the core came out, it was about eighteen inches long by four inches in diameter, of the most delicious, nutty, white substance, as nice as the finest filbert. About 200 or 300 yards above where we crossed the River Leichardt, we saw a recently-marked tree on the west bank of the river, branded broad-arrow over L over 10; it cannot have been done many weeks. We crossed just at the junction of a considerable creek, joining the river from the west. I cannot find that it has any name. We crossed the river without much difficulty, a few trees or saplings had to be cut down, and the hind-wheels had to be locked. The bed is gravelly; and on the east side, where we crossed, there appears to be a bed of gravel, good for ballast. We had dinner after crossing, just exactly at noon. Our dinner camping-place was one of the prettiest I have seen. We then had a fresh team, and as the road was good, and we had the mailman's track, we came along at a rattling pace, keeping pretty well within sight of the river all the way. The country is flat, or very slightly undulating ; well grassed and lightly-timbered plains—bloodwood, bauhinea, what I call blackbutt or Moreton Bay ash, and whitegum. There are ant-hills. I find the whitegum has been used for buildings ; but the white-ants destroy them. There is plenty of stone in and near the river. We came about fifteen miles after dinner, and then camped in a very pretty part of the river, where there is abundance of feed and splendid sheets of water. We met the mailboy on our way ; he was going to Cloncurry. I gave him telegrams for the Colonial Secretary and Mr. Collier. The river, where we crossed it the second time, is pretty much the same as where we crossed it first, as far as the cost of a bridge is concerned; about the same provision would have to be made for water. On our way, to-day, down the east side of the river we did not meet with any troublesome creeks, but we crossed several curious places called "breakaways," where some scores of acres of the soil, to a depth of three or four feet, had been washed away as clean as if it had been excavated, looking very much like irregular side-cutting pits. The grass-seeds have not been quite so bad on this side of the river. Barometer at 5 p.m., 140 ft. or 29.47. During the afternoon we saw more horses and cattle than we have seen since we left Mount Cornish, and all in first-rate condition. This has been a most satisfactory day as far as progress goes, and all looked well when we camped, but the pack-horses were a very long time coming, and when they came we found Tappenden, the groom, had had a nasty fall and roll over with his horse. He was seriously hurt. It occurred about ten miles from the camping-place, and he could only proceed very slowly, Mr. Baynes accompanying him. Shakelton went at once to look for him, it was now dark (7 p.m.), but came back about 8.30 p.m. unsuccessful. Mr. Hann then went, and at 9.30 p.m. they all came in. Tappenden was very much shaken and bruised, but nothing broken; we put him to bed. Shortly after a row commenced between Mr. Hann and Mr. Wyatt; it originated about the fetching of water from the creek. Mr. Hann has an unfortunate temper, and completely lost it so far as to challenge Mr. Wyatt to see who was master by fighting, and, failing to get Mr. Wyatt to respond, he struck him in the face two or three times. This was most unjustifiable; Mr. Wyatt behaved with a great deal of forbearance. What previous provocation there may have been I do not know; evidently something has been rankling; but nothing could justify the blows. Mr. Hann is a most indefatigable, energetic, and useful man; but his temper is a perfect curse, and, I think, would soon lead to mutiny in any camp. It is such a pity.

TUESDAY, APRIL 19.—Tappenden has had a pretty good night, but is unable to proceed on horseback, so he will go in my place in the waggon, and I will ride. We got away about 7.45 a.m., without any fuss. Mr. Wyatt spoke to me about last night. I expressed my regret, but could scarcely blame him. Mr. Hann also spoke to me, and I told him he was very wrong to strike Mr. Wyatt. The two men are totally different in all respects. We came through some splendid country all the way to Kerr's Station (Camilroy), about twenty miles to-day, fine undulating plains, lightly-timbered, and most luxuriously-grassed. Very few creeks or watercourses; those that there are are well defined; only one creek of considerable size, about five miles from camp. This we crossed at its junction with the Liechardt; in fact, we crossed it in the river, or rather rounded its mouth. We saw a lot of cattle, all well to do, sleek and happy. We passed a lot of pretty lagoons, but they will soon be dry. The river, wherever we saw it, was very beautiful, and the country very fine. When we stopped for dinner I was under the impression that we were close to the station (Camilroy). I afterwards found we had fully five miles to travel. We found Mr. Curr at home, and I stopped with him during the remainder of the afternoon. Mr. Hann came to dinner. Mr. Curr is very doubtful about the Point Parker harbour, and speaks of the enormous rivers to be crossed to get there—the Liechardt, the Nicholson, the Gregory, &c. He thinks a harbour could be found or made east of the Liechardt. We camped near the station. He says there will be no difficulty in reaching Brodie's (thirty miles) to-morrow, as the road is good. He confirms what Mr. Hann and others have said about the large rivers that would have to be crossed if the line went down west of the Liechardt. The country has been splendid all day : abundance of feed, with a good deal of Mitchell grass. He says cattle get fat in spinifex country. We saw plenty of stone, and some sleepers might be got (bloodwood, &c.). Riding on horseback does not agree with me very well—I became very tired. Tappenden is evidently badly shaken. The flies are dreadful. Did not read barometer ; forgot it. Camp 48.

WEDNESDAY, APRIL 20. — Barometer, 000 ft. or 29.36 ; thermometer, 98°. Put down barometer 1000 ft., as there will be further fall. Got away soon after 8 a.m. Tappenden going in the waggon and I on horseback, but the road was very rough, and he could not stand the jolting, so we had to change. Mr. Curr accompanied us for five or six miles, to show us the way. At about ten miles we came to the junction of the Gunpowder Creek, which comes in on the west side of the Leichardt. Mr. Hann and I rode across to see it. It is just as large as, and very much like, the Leichardt. I saw my first crocodile in the river, but very little of him, only his eyes and forehead. He was very shy. We camped here for dinner about noon. The country we have come over to-day is a magnificent open plain for many miles, splendidly grassed ; then, when nearing the river, lightly-timbered with bloodwood, coolebar, &c., but not of any useful size. The feed continues very abundant, and every now and then there are splendid lagoons, some of which hold their water nearly all the year. Plenty of water in the river and Gunpowder Creek. After dinner we passed through the same magnificent country for about twenty miles, and we reached Brodie's Station about 5 p.m. Taking into consideration the richness of the soil, the abundance and goodness of the feed, the grateful shadow of the timber, and, above all, the splendid supply of water in the Leichardt, supplemented for a great part of the way by lagoons extending a considerable distance inland, I have not seen any better land for agricultural or pastoral purposes in any of the Australian colonies. In Mr. Brodie's house there are a good many whitegum slabs, and they are looking remarkably well. Discovered a break in the fore-carriage of the waggon. Decided to remain here to-morrow and repair it, although it will occupy some hours. Referring to the unfortunate stockman recently murdered by the blacks, Mr. Curr told me that the stockman and a black boy were hunting for stray cattle, when they came upon the blacks' trail, which they followed to their camp, then they drove away the blackmen and kept possession of the gins, with whom they remained in the camp. Presently the stockman fell asleep, one of the gins stole his revolver, and gave the signal to the blackmen, who came and put a spear through both of his thighs, pinning him to earth, and then beating out his brains with their nullahs. They then cleared out. This is the boy's version, but he did not report the matter for four days. It seems the stockman had been thrashing him for some days, and it is thought he may have had his revenge. Mr. Curr told me that he and others had pursued the blacks and shot five ; that the police were coming to give them a further "dressing," as that was the only thing they understood. It seems hard upon a man, whose gin or child is stolen, to shoot him because he objects, but I believe there is no help for it except speedy and ostensible annihilation. The conduct of many of the whites to the blacks is simply disgraceful. The name of Brodie's Station is Lorraine. It is about sixty miles above Floraville, and on the Leichardt River. They were exceedingly hospitable, and we had all our meals with them. This is Camp 49.

THURSDAY, APRIL 21.—Barometer, 940 ft. or 29.70 ; thermometer, 71°. The whole of the day, up to nearly 4 p.m., was occupied in putting a new fore-carriage to the waggon. Our best hand, Tappenden, being disabled, not able to do anything beyond looking on and giving his advice, the work fell upon Mr. Hann and Shakelton. I think they have made a splendid job of it, and am very much in hopes we shall reach our destination without further repairs of any consequence. Tappenden is so much bruised and shaken that it was deemed best he should spell for a few days at Mr. Brodie's and join us at Burketown or Normanton. We made a start at 4 p.m., and did about seven or eight miles through the same splendid-looking country as for the last two or three days. Then we camped on the banks of the Leichardt, amid splendid pasture and close to a very fair water-hole. The country is lightly timbered with box, bauhinea, bloodwood, &c. I wrote Walsh this morning, and sent telegrams to Colonial Secretary and Collier, Mr. Brodie taking them to Normanton. The chances are that they will reach their destination before some of those sent from other places several days earlier. This is Camp 50 ; barometer, 1060 ft., or 29.57, at 5.30 p.m. This is about 100 yards below marked tree, broad arrow over L over 85, on west bank of river, recently cut. The barometer shows that the table-land where we are camped is about 55 feet above the bed of the river, or above the water, which is three feet deep; and the highest flood has been two feet over the table-land. The barometer has fallen considerably during the day. There is lightning and thunder, and indications of rain. Camp 50.

FRIDAY, APRIL 22.—Barometer, 950 ft., or 29.70. It began to rain heavily about midnight, and continued during the night, but the barometer is higher this morning. The weather continued very uncertain during the whole day, with frequent showers, and little sunshine. At one time we were almost certain to make a start after dinner, but the rain came on again, and, as everybody's clothes were wet, it was decided to dry them as well as possible and make a start in the morning. There was plenty to do in writing up notes, distances, &c. It appears that up to the present time we have travelled 1123 miles since we left Roma, on 14th January, and, as there have been fifty camps, we must have travelled fifty days, or an average, when travelling, of nearly 22.5 miles per diem. But, in addition to this, I travelled by coach and train to Rockhampton and back, about 850 miles. The delay to-day has enabled us to put together a lot of information about the camps, distances, &c. My eyes are better, but still itching most painfully. The barometer has been falling ever since the morning, and in the evening read 1040 ft., or 29.58.

SATURDAY, APRIL 23.—Barometer, 1040 ft., or 29.58; thermometer, 68°. Heard a noise in the river about 3 a.m. Got out, and found a flood coming down. It continued to increase, and by daylight there was a very considerable stream running rapidly. This was just twenty-four hours after we had the heaviest rain here. There must have been much heavier rain up stream, and we have reason to fear it may impede our crossing at Floraville. My eyes are much better this morning. The river was still rising when we got away at 8 a.m. The country much the same as for the last few days ; fine soil, fine feed, light timber (bloodwood, bauhinea, &c.) After about seven or eight miles, less timber, the soil changes from brown to darker, which the rain has made very boggy. Horses down to fetlock at every step, and soon began to show distress. We had to put in the strong team at 10.40 a.m. At 10.20 a.m., or about ten miles from Camp 50, we saw a marked tree, broad arrow over L over 95, some distance back from east bank of river. This boggy ground is sadly impeding our progress, and knocking all our calculations upside down. The pack-horses are able to keep up pretty well. Even the

G

strong team nearly caused us to lose faith in them. They had a grand struggle to start with; the bog did not continue far, but presently we came to a worse one, and one of the wheelers went down. This brought us to a long stand-still; but shortly we got away again, only to go a mile or two, when the off-wheels went down to the axles like a shot. The off-wheeler and the two leaders all went down, completely beaten. It took us a long time to get right. It is very unfortunate, for we were heading the flood, and but for these delays should have crossed the Leichhardt in safety. Now no one knows when we may. But we had our dinner, notwithstanding. We had proceeded with varying luck as to the state of the ground for a few miles, still retaining the "emergencies" for fear of accident, when Mr. Hann, who had been ahead exploring, returned to say that he had come upon a creek which he had entirely forgotten. It was not a large one, but one extremely difficult to approach. After a very careful examination up and down for about an hour, it was decided it was impossible to cross it. The only thing now was either to remain until the ground had dried, probably two or three days, or to go round the mouth of the creek after it had joined the river. The flood was fast following us down the river, but we were still a little ahead of it, and it was decided to adopt the latter course and round the mouth of the creek. We were not unmindful of the danger of adopting this course, for we might be bogged in the river and the flood might come upon us; but we were all anxious to get on, and so started. The banks of the river are rather steep, but we got down safely and then found the sand was extremely difficult to travel over. Our horses, too, staunch and true as we knew them to be, had been a little bit cowed by the two nasty boggings they had already had. The creek empties itself into a billabong of the river whose effluence is about half a mile above, and its confluence half a mile below the junction of the creek. Its bed is higher than the bed of the river by several feet. Its effluence has a bar still higher than its bed, and also much higher than the confluence. There is an island, pretty well timbered, between it and the river, several feet, say twenty feet, above the billabong, but sometimes covered in floods to a considerable depth. The water had not yet commenced to come in at the effluence, but was just beginning to back in at the confluence, and the river still rising. The danger was imminent if anything happened to delay us whilst we were in the bed of the billabong; the water would be upon us, and we must leave the waggon and save ourselves. There was not a tree, or stump, or stone to which the waggon could be moored, and it must, if the water rose sufficiently high, be carried helplessly down the stream. Presently our worst fears seemed about to be realised, for, on crossing some water about eighteen inches deep, by swerving a very little from our intended course, our waggon and horses became most terribly bogged. In addition to the eighteen inches of water, the wheels and the horses' legs had gone another eighteen inches into the sand. After struggling, kicking and plunging in a most distressing manner for several minutes, all four were at last down in the sludge, entangled in their own harness and each other's, one apparently hopeless mass of confusion; and all this time the flood increasing. It was a most critical moment, and not a second to be lost. All hands were mustered round, and somehow the horses were soon got free, but how I don't know. The harness was terribly smashed and miserably entangled. At last the leaders were connected with the end of the pole, some of the sand was cleared away in front of the wheels, and then the two horses who seemed to know what was required from them, and four willing hands at the wheels, altogether, with a long pull and a strong pull, landed the waggon safely on harder ground. The other horses were soon fixed, and the waggon was got on safe ground just as the flood was beginning to back in furiously from below.

Half an hour more would have left the waggon helpless in the bed of the billa-bong until it was washed away. The danger was now over, but there was yet another difficulty to surmount, viz., to reach the tableland. A great deal of clearing and excavation had to be done, as the bank was precipitous; the sun, too, nearly down, and the twilights in the tropics well-known to be ex-tremely short, miserably short on occasions like this. However, our brave horses took us to our camping-place safely, to our great relief and surprise. This has been a most troublesome and exciting day, one that will not readily be forgotten by those that were present, but will scarcely be realised by any-one else. The evening was fine, and there is reason to hope we may get on better to-morrow, as there are no more creeks between this and Floraville; but the flood has sadly interfered with our intended progress. However, "all's well that ends well." Camp 51.

SUNDAY, APRIL 24.—Barometer, 1000 ft., or 29.24; thermometer, 69'. There was a good deal of "fitting and contriving," hammering, stitching, and riveting to be done this morning, and we did not get away before 8.40 a.m. The ground was so boggy that we were afraid to go beyond a walk, which means two miles an hour, and thus we went on for hours, the weather fine, and the country looking beautiful. But for the recent rain, we might be bowling along seven or eight miles an hour instead of crawling as we are. At 11.15 a.m. we came to a halt, as there was a creek ahead, and it was uncertain whether we could cross it without going a mile or two up to head it or nearly so. Mr. Hann and the cook went away to explore. They came back in about an hour, having found a crossing-place near the junction, after exploring several miles up without success. But we had to put in the "emergencies." The creek proved far less formidable than anticipated. There was no water in it, no bogging; only loose sand, which let the horses and wheels down about eighteen inches; the "going in" and "coming out" were pretty steep, but the horses went straight ahead, and within five minutes we were over. There are some trees here, such as we have not seen before—broad leaves, 6 in. x 4 in., and plenty of them; the most perfect shade of any I have seen. Mr. Hann says it is the "pear tree," but he is not certain. After crossing two or three more creeks and one open plain, we came to a fine lagoon, and camped for dinner at 1.45 p.m. We are promised better country this afternoon, and we want it badly, to make up for loss of time and patience during the last few days. We saw some india-rubber trees on our way; small, scrubby-looking things; no foliage; something like a blackthorn just beginning to bud; if cut, a milky fluid comes out, but very slowly. I suppose this is the fluid from which the rubber is made. The lagoon on which we had dinner is supplied during floods from the Leichardt. It is dry in some seasons. There is very little timber. Some sleepers may be got on the river, but I don't think the ti-tree can be relied on to furnish any large quantity. There are some good-sized bloodwood trees, but I notice that many of them are hollow. No stone to be seen for the last few days, but the nature of the soil indicates some underneath. We had a new team after dinner, but did not, for a mile or two, come upon anything that could be called trotting ground, and by that time our leaders were pumped. We put in a fresh four, and then we made good progress. The country still continues much the same as for the last few days, a little less timber and occasional indications of sandstone. It is said there a great many crocodiles in the river near here. I was called to see two or three, but they turned out to be logs or stones. At last, however, I saw a most unmistakeable one on a boulder; saw him move, crawl into the water, and disappear all but his head. This Mr. Hann hit with a bullet from the rifle, but bullets do not hurt them, except in some soft place behind the arm. We reached a good camping-place on a nice lagoon about two miles off the river. We have come about eighteen or nineteen miles to-day. Our progress is slow and very aggravating, but we must reach Floraville to-morrow, and cross the river if possible. This camp is 52. Barometer, 1050 ft., or 29.57, at 5 p.m. There is some rather soft conglomerate stone here; it would make very fair ballast, but there is no lack of ballast anywhere.

MONDAY, APRIL 25.—Barometer, 970 ft., or 29.67; thermometer, 63°. We started at 8.5 a.m., and, after about two miles, came to the river. We followed it for about two miles, and then came upon some rocks of conglomerate on each side of the stream. I think foundations might be formed on them, one on each side of the main stream, not more than sixty yards apart. These rocks or cliffs are about twenty feet above the level of the water. There is now a slight flood in the river. In very ordinary floods, no doubt, the cliffs are under water. They extend back level from the river for a considerable distance, and are sometimes flooded to a depth of twenty or thirty feet. This place is about twenty miles above Floraville, and a little above Shingle Creek, which comes in from the eastward. It is not a bad site for a railway bridge. We found some very hard stone here, like cornelian. Shortly afterwards we came to Shingle Creek, whose bed is one mass of rocky boulders, enough to try the remaining strength of the waggon. Then for miles, rather for hours, we were travelling the roughest of plains; good soil enough, no doubt, but jolting enough to shake the Christianity out of any man. Then we went down into some broken country, and for a few miles it was harder and a little smoother. The sorghum grass covers the plain, and it is like driving through a sparsely-cropped wheatfield. Some of it is eight feet high, and it is most irritating to the poor horses. There is also a sprinkling of Mitchell grass. We stopped for dinner at 12.45, and here a carpet-snake six feet eight inches long was killed. After dinner, with a fresh team, we had more than the usual tussle, but the ground is enough to try the tempers of man and horse. I am in constant fear of my neck being dislocated by my head coming so frequently into violent contact with one of the uprights of the cover of the waggon. We went on with varying success (mostly ill-success as far as the ground went) until half-past four, and then our team was completely pumped. We had to let them out and fall back upon the " emergencies," having five miles more to reach Floraville, and the ground still uncertain. Just before we came to the camping ground, with the sun nearly down, our connection between the pole and the leaders was broken. We soon repaired, temporarily, with a chain, and got in safely just before sundown. It has occurred to me that if the troublesome sorghum grass through which we have been driving and swearing all day be allied to the sugar-cane, why should not the sugar-cane grow luxuriantly also? It appears, after all, we have done twenty-four miles to-day, but the incessant jolting has been trying. Camp 53. Barometer, 940 ft., or 29.70, at 6 p.m.

TUESDAY, APRIL 26.—Barometer, 870 ft., or 29.77; thermometer, 72°, at 6 a.m. Immediately after breakfast, Mr. Hann and I started on horseback to see something of the big fifteen-mile waterhole, &c. We reached the river at tree broad arrow over L over 143 in about two miles, and followed it about two or three miles down to the falls. The waterhole is a splendid sheet of water, about a quarter or one-third of a mile wide. The falls are very fine, though not " Niagaras." There are two, one on each side of the river, with a ridge between them. It is a sort of double fall; first a ledge of five or six feet; then a flat, perhaps twenty feet; then the main fall, about twenty

feet. The formation is a hard conglomorate, in layers or strata, varying from
1 ft. to 12 ft. Some of it is very close-grained, almost like granite. The main
watercourse is tolerably well defined, but in floods it extends a long distance
on either side; I think a mile of waterway altogether. 300 feet, in ten
30-ft. spans, and the remainder in 15-ft. spans, averaging about 15 ft. high,
ought to suffice, and should not cost more than £33,000. This is mere con-
jecture. The river Landsborough (? Alexandra) joins the Leichardt on the
eastern side about half a mile below the falls. From the Leichardt Falls we
went to the point at which we were to cross the Landsborough, and where the
party was waiting for us. About a mile from its junction with the Leichardt,
we crossed immediately above a very pretty waterfall, not so extensive as the
Leichardt Falls, but much prettier, having a beautiful basin, and a sort of
cavity underneath the ledge, over which the water falls, where one could stand
and look through the falling water. The bed of the river where we crossed
is all rock, the same sort of close-grained conglomerate as on the Leichardt.
We were all across at 9.30 a.m., and had then four miles (bush miles) to go
before reaching the Leichardt crossing, which was to be the event of the
day. We got to it about 11.10 a.m. Mr. Hann rode across it. It looked
very exciting. On his return he got upset; his horse fell, and he was up to
his shoulders in the middle of the stream; but he came out all right. It was
evident the water must come into the waggon, so everything perishable, or
liable to be injured by water, was taken out and stowed and lashed upon
saplings laid from side-to-side of the waggon. When all was ready we
ventured. I rode the cook's horse; he and the driver went in the waggon;
Mr. Hann and the black boy, "Jemmy," leading the leaders and wheelers.
The waggon and horses had not actually to swim, but very near it. In fact,
as I struggled along behind the waggon, I saw sometimes the hind-wheels
did not revolve, although the waggon was going on. Now a wheeler went
down, then a leader, and once or twice the case seemed hopeless; but, by
dint of coolness and perseverance, we landed with no greater injury than a
good soaking, and slight damage to some unimportant papers. We got the
waggon safely landed on high ground, and then came the excitement of seeing
the pack-horses cross. It really was a pretty sight; only one casualty—Mr.
Baynes and his horse disappeared, though a thorough soaking was all the
damage, and dinner seemed soon to repair it all. We started again soon
after 1 p.m., all congratulating themselves on their good luck in having
crossed such a flood in safety; another inch or two must have cooked us.
As it was, some damage was done to one or two of my books. There was
thunder knocking about; and, during the afternoon, we saw some heavy rain
falling up the Leichardt. Mr. Hann left us soon after crossing the river, to
search for some camp for which he had letters. We went on a few miles
until we reached and crossed Bothwick's Creek, where we camped. The
country from Camp 53 to the Landsborough (? Alexandra), and from the
Landsborough to the Leichardt, is very much the same as yesterday's; a
little less "sorghum," and a little more "Mitchell" grass. After crossing the
river as far as the creek (Bothwick's) it is a fine, open, well-grassed plain,
good soil, but no timber. Half a mile before reaching the camp it came on
to rain heavily. Camp 54. Barometer, 890 ft., or 29.75, at 5 p.m.

WEDNESDAY, APRIL 27.—Barometer, 820 ft., or 29.73; thermometer, 75,
at 7 a.m. We got a good start at 7.35 a.m. with the strong team, but our
progress was miserably slow. The rain last night, acting upon the rich brown
soil, made the wheels clog frightfully. They had to be cleaned every 100 or
200 yards for the first three or four hours; and, at times, we were scarcely
making one mile an hour. The ground soon dried up and got less sticky;

by-and-bye no more clearing of the wheels was required. It was, however, a quarter-past one when we stopped for dinner, and we could not have done at all more than ten miles. We have come over some very rich soil, with some good pasturage, Mitchell grass, &c.; no sorghum until the last mile before dinner. We had dinner at a fine waterhole on Millar's Creek, close to the track, about two miles before coming to Millar's Water-hole. Our strong team was pretty well knocked up. It has been a very trying morning for them; constant jolting and nagging. After dinner we had a change of team, and great bother at starting. Presently I changed places with "Pagan," the black boy, partly because he was lighter than I for the waggon, partly because I had had shaking enough in the forenoon. I rode off with the pack-horses. We went over some dreadfully jolty ground, but this did not much affect the pack-horses. We passed over a lot of fine country all day; splendid, rich, open plains, well grassed, and the pasture good. We, with the pack-horses, reached the camping ground on Crooked Creek at 5.5 p.m.; a capital camping-place, plenty of feed and water, and the shadow of the peartree. At 6 p.m., the waggon was heard approaching, but Crooked Creek had to be crossed, and it proved a very awkward one. The horses were beaten, and would not pull steadily. Constant fits of jerking kept breaking the harness. At last it was decided to fall back upon the " emergencies," and they took the waggon straight to camp. It was now 7 p.m., and just about dark. We have come about nineteen miles to-day, over very bad roads, and hope to get pretty near to Burketown to-morrow, as the most liberal estimate makes it only about twenty-five miles. This is Camp No. 55. Barometer at 7 p.m., 870 ft., or 29.79.

THURSDAY, APRIL 28.—Barometer, 840 ft., or 29.81 ; thermometer, 77°. Lots of repairs had to be attended to, and the weather looked very unpromising. Got away at 9 a.m. with a police team evidently worth nothing. With a great deal of persuasion and coaxing, they were got on half a mile ; then the leaders were changed, but without success, and the emergencies had to go in. I rode on with the pack-horses, "Pagan" taking my place in the waggon. We came across some fine open plains, then some light timber. Showers have evidently fallen at irregular intervals, and the wheels had to be cleaned many times ; but, notwithstanding the delays, we were all at Harris's Creek at 2 p.m., with only a short afternoon before us, as we cannot cross the Albert River to-day, and there is no water to enable us to camp near to it. The waggon is getting very shaky, and it seems at times doubtful whether or not it will take me to Point Parker. We shall have to spend some time at Burketown to-morrow for repairs. I rode again in the afternoon. A fresh team was put in. The road was rough, and there was a dreadful game at starting, but once the collars warmed they were all right. I examined the water about half a mile down the creek from where we had dinner, and it was quite salt. The tide was up. We came about seven miles across some very rough plains, very rich, and plenty of feed ; some scrubby trees, no good ; no stone on surface. We camped on what is called the Lakes, about three miles east of Burketown. We could have gone on to the Albert River, but there is no fresh water there. Mr. Hann went on to make enquiries about the tide for to-morrow. We found some very mild blacks at the Lakes, two gins, mother and daughter, and the latter had a little piccaninny. They were pleased to be noticed. Camp 56. Barometer, 850 ft., or 29.80, at 5 p.m. From the flood-marks in all the creeks, and the general appearance of the country, it looks very much as if it was all submerged from the Leichardt to the Albert in times of very ordinary floods ; and the flood-marks at " Harris's Creek," with the close proximity of the tides from the

Gulf, help towards this conclusion. It will be necessary to keep the line higher up, crossing the Leichardt at the rocky place about twenty miles above Floraville. Mr. Hann returned to the camp about 9 p.m., having lost his way for some time and only found it by "coo-ee-ing," showing lights, &c. There came with him a Mr. Sutherland and a Mr. Child. The latter, on his way to Normanton, has offered to take letters, &c., to the post-office for us. They gave a very exciting account of the crossing of the Albert. It seems the alligators muster in great force there, and have to be frightened away by yelling and loud language. It is said they are of enormous size, and do not hesitate to attack a horse; in fact, one horse is said even now to bear the marks of an alligator's upper teeth on his back and his lower teeth on his belly. I have not yet seen him. We cannot, on account of the tide, cross the river before to-morrow evening. Camp 56.

FRIDAY, APRIL 29. — Barometer, 820 ft., or 29.84 ; thermometer, 73°. Repairs occupied about one and a-half hours. Got away about 9 a.m., having first written some letters and telegrams to send to Normanton by Mr. Child. Although we may have a big day's work, we have not far to go. Our black friends had remained close alongside us all night. We had a composite team to start with ; two "emergencies" for leaders, and two troublesome quick-steppers for the pole. A worse selection could scarcely have been made. You cannot expect oil and vinegar to unite ; just the sort of thing to spoil both teams. However, we soon changed it, and got two more suitable leaders, and thus we got to a lagoon or billabong about one mile from the Albert River. (We had come over some good-looking, grassy, open plains, rather jolty.) Here we left the party ; Mr. Hann, Mr. Wyatt, Mr. Baynes, and I riding on to explore the river ; a noble river it is. The tide was up. It was about noon, and would not be at its lowest until 6 p.m. Does it not follow that there are two tides per diem? We rode up and down the river. Saw three alligators. Examined the crossing-place to see what had to be done in the way of approaches ; then returned to camp and had dinner. At about 2.30 p.m. we started, and reached the river in about half an hour. A good deal of excavation had to be done to make a way for getting into the river. The getting out seemed to be all right. Lots of branches were cut down and laid in the river, and upon some of the excavated material ; then other material was excavated and shifted on to the branches. At last the "in-going" was finished, and Mr. Hann and "Pagan" rode across. What seemed to be a way out proved to be something like a boggy quicksand, so another had to be formed. This was close by, and a little excavation and a few bushes made it all right. The tide was still falling, but the sun was setting. We knew that he must and would set, and at his appointed time. We knew also of the short, unpitying twilight, so we had to " go in." We had been cautioned that we must yell and make a great noise to keep away the alligators. Everybody seemed to yell instinctively, but I don't think it was necessary to keep away the alligators, which have too much good sense to come near "when two or three are gathered together," and prefer to sneak their prey—come suddenly upon a solitary, unsuspecting individual. I have said we had to " go in," and, after a little struggling and persuasion, the four emergencies got in, taking the waggon with them, with a great plunge and a splash ; but they had not gone far when the horses got " mixed " somehow, and the harness entangled, some of which in the struggle broke. The water was up to the men's armpits. At one time it seemed certain that the waggon would have to be abandoned in the river, for the night at least, if not alto-gether, as the tide would rise about three feet ; but, by indomitable perseve-rance, things were made straight, and the splendid team took the waggon across and up the steep bank in a most perfect way, to the admiration of all, including the entire population of Burketown, numbering seven all told. The pack-horses now had to be dealt with, and this was even more difficult than the waggon. They could not be persuaded to "go in." The yelling was terrific, and this time accompanied by such language as only, I hope, Queenslanders can use. The oaths of the Queenslander are peculiar, and to a great extent senseless. He collects all the obscene and profane words he can possibly think of; jumbles them together anyway, utterly regardless of their meaning, if, indeed, in his own mind they have any meaning, and then belches them forth with the most savage ferocity, accompanied by the most grotesque and senseless gesticulation. At last they were all "got in," and all struggled out without any accident. Of course, everybody was thoroughly drenched to the skin, from the crown of his head to the soles of his boots. We got safely to a middling camping place, No. 57, near the township site, just after dark. The people were most kind in cheering and assisting us in our difficulties. There are not many inhabitants. I think the township had to be abandoned in 1872, in consequence of being unhealthy; and now I think the cemetery is more thickly populated than the township, so that if a man dies here he literally joins the great majority of his own little world and time, irrespective of the great world at large and all time. We are promised some assistance about repairs in the morning. Accounts of the country between this and Point Parker are very conflicting, and given with great hesitation; some say fifty miles, two days easy riding, some say sixty miles, straight, but that you may have to go 260 miles to reach it. The only thing is to try, with determination to succeed. The weather is now in our favor, and I think we shall go forward in the morning with every confidence that less than a week will see us at Point Parker. All through the difficulty in crossing the Albert River Mr. Hann was the leading spirit. He led the "forlorn hope," and he led the "language" in all its branches most effectually and forcibly. When the end is success, then one can forget the means.

SATURDAY, APRIL 30.—Barometer, 770 ft., or 29·89; thermometer, 68°, at 6 a.m. Camp 57. Immediately after breakfast, we shifted the camp about a mile, to the ruins of the old boiling-down establishment, because there was plenty of wood and water there, and it was more convenient for doing the repairs. A good man at this sort of work was got from the township—a carpenter, called Shackell. We left the matter in his hands and Shakelton's (our driver), and Mr. Hann and I went on an exploring expedition. We started soon after 9, and followed a track which leads to the Old Brook Station, about six miles; then turned to the left, and in about two or three miles struck the Albert River, about four miles below the junction of the Barclay River and the Brook Creek—these two, in fact, forming the Albert River. It is a splendid stream at this point. There is no tide. The water is quite fresh. The river is well defined, but the floods extend a long distance on either side. A line might come across the Leichardt, twenty miles above Floraville, to some point about here, and continue on straight, crossing the Nicholson where an important effluent leaves it, and still straight twenty-four miles further, bearing about N.N.W.; thence due north to Point Parker, about thirty-two miles. This is, of course, mere guesswork, but it is the route I propose to examine. From the Albert we rode across to the Nicholson, stopping at a lagoon to have our dinner on the way. It is fair country to travel over, but when we struck the Nicholson, which is another splendid river, larger than the Albert, we followed it down the stream about a couple of miles, all the way along a splendid water-hole, to where an effluent leaves it and goes direct to the sea without joining again. At this point, which is the lower end of the water-hole, we found a splendid crossing-place, rather wide, but not up to the horses' girths. We decided to cross here to-morrow, if possible. There are crocodiles about here. We saw distinct footprints, apparently not many minutes' old, on the sand. We must have disturbed them. There is some fine timber here (bloodwood, &c.) We have ridden over some very fine plains, well stocked with good grass. On our way back we rode over five or six miles of terribly rough, jolty plains, called the Devil-Devil country, and no wonder, as it is fit for only the devil to travel over. The roughness is frightful, and I think occasioned by frequent floods submerging it. By-and-bye we came to Woods' Lake, a very fine sheet of permanent fresh water. It is five or six miles long, and from a quarter to half-a-mile wide, well stocked with wild fowl, and here, as it was evening, we saw thousands of sleek, well-to-do cattle. We unfortunately took the north side of the lake, and had to ride several extra miles to get round it and its swampy adjuncts. We called at Boydale's hut and had some tea, not getting to the camp before 8 p.m. I shall call this camp 57A. We found the waggon thoroughly repaired, and I hope now fit for Point Parker. I was very tired after so many hours' riding, but I think the information I have gained well worth the trouble.

SUNDAY, MAY 1.—Barometer, 780 ft., or 29.88 ; thermometer, 71°, at 6 a.m. An examination of the waggon leads us to hope that it is all right. On looking at the tracing, the route to Point Parker, after crossing the Nicholson, as suggested yesterday, seems to be N.N.W. about twenty-four miles ; thence N. thirty-two miles. The direct course would be N. by W., and the distance 54 miles, but this would run into the Gulf many miles before reaching Point Parker. Some rations (sugar) had to be procured, but we got away very shortly after 9 a.m. ; and, as the horses had a track, they came along at a spanking pace after they once got their shoulders warm and their tempers set to rights. This track leads to what was the Brook Station, and is not our direct course, but it saves time to go out of the way for a beaten track. We followed this track for about eight miles, and then turned at right-angles for the Nicholson River, about five miles, the first three and a-half of which was over "Devil-Devil" country. Then we came to the river timber and scrub, bad, but not so bad as the other. A Mr. Wetherall and Mr. Boydale came with us to, and across, the river. The crossing was easy and successful. Then we had dinner. Safe over river at 2 p.m. After dinner we came over some very beautiful-looking country for about eight miles, lightly timbered, high and dry, well-grassed, and undulating, with a few unimportant but well-defined creeks, where there is abundance of ballast in the form of a sort of ironstone conglomerate. There is a good sprinkling of useful bloodwood. We found a nice convenient camping-place about twenty-two miles from Burketown, and consider we have done well to-day. The arrangement of the rivers here is peculiar. About sixteen miles above Burketown, Brook Creek and Buckley River unite, and form the Albert River, which goes into the sea ; but the Brook Creek comes out of the Gregory, and Buckley River and Brook Creek separate before they join to make the Albert. Then the Gregory goes into the Nicholson, and the Nicholson goes into the sea. So that the Albert, the Brook Creek, the Nicholson, and the Gulf form an island about 100 miles by ten or twelve miles. I am not very clear about all this, and had better, perhaps, have said nothing ; but it will refresh my memory if it serve nobody else. The weather looks fine, and everything appears promising for an early finish to our labors. I have been on horseback all day again to-day, but have nearly had enough of it, and now the country is getting easier. At Camp 58 at 6 p.m. Barometer, 860 ft., or 29.78. Mr. Hann has verified what I have said about the rivers and the island.

II

MONDAY, MAY 2.—Barometer, 825 ft., or 29.32; thermometer, 69°, at 6 a.m. We made a good start at 7.40 a.m., having to-day taken the precaution to distribute the arms and ammunition. We came at a grand pace over level or slightly undulating country, thinly timbered, well grassed, but not rich soil, and occasionally some scrub that required a little clearing work with the tomahawk. We went on in a N.W. direction until 10.10 a.m., and must have done thirteen or fourteen miles. We now came upon a fine sheet of water, which we think must be what Mr. Edkins called Moonlight Creek. The water is fresh, or very nearly so. We passed some ironstone conglomerate on our way. I do not think there is any timber that will be of service for the railway. The country is exceedingly pretty, but I do not think the soil is very rich. We had to put in the "emergencies" to cross the creek. The pack-horses are keeping well up, and we seem likely to do a good day's work. We were safe across at 10.45 a.m., and then we had our lunch. The creek appears to divide into two just below where we crossed—one branch (the largest) going east, the other nearly north. There are some very beautiful shady, but scrubby, trees here; they are a little too thick to admit of our making rapid progress. Our course after dinner (we started precisely at noon) was, as near as we could shape it, N. by W. After an hour and a half of fair trotting ground, we came upon a pretty water-hole, standing east and west. We had to round it, which took us nearly a mile out of our direct course. Soon after this we got into a most inconvenient scrub; a lot of cutting-down had to be done, and for more than half-an-hour our progress was very slow. Here, too, my poor tilt came to grief, and had to be abandoned. The branch of a tree and four horses were too strong for the tilt, so the weaker had to give way; it was hopelessly smashed. After a very short consultation, it was decided to dismantle the frame, and, leaving it behind, to take the canvas with us. This was soon done, and presently we got into better country. We passed several lagoons, one remarkably fine one. The waters are brackish. We crossed a considerable inlet of the Gulf, where the tide evidently comes everyday. When we got there (3 p.m.) it was low water, so there was no difficulty in crossing. We soon afterwards camped for the night, although it was only 4 p.m. We must have got over twenty-five miles to-day, and all are pretty tired. Mr. Hann rode on for a few miles to explore. He seems never to be tired. This is camp 59. Barometer, 910 ft., or 29.73, at 5 p.m. We have passed several patches of conglomerate stone to-day, but not much timber suitable for a railway. The country is low, and no doubt flooded at times, but not impracticable, as we have come. No doubt better could be found. We saw enormous ant-hills to-day, seven, eight and ten feet high, and I measured one monster fifteen feet high, and the same in diameter; they are conical, but with rounded tops. All the very large ones seem to be forsaken.

TUESDAY, MAY 3 —Barometer, 850 ft., or 29.80; thermometer, 76°. Mr. Hann's exploration last evening proved to be of the greatest value; it showed that we cannot go straight ahead or we should get into a dreadful patch of scrub, which would delay us for hours. We have to turn sharp to the right (east) for about four miles, then north over some fine salt plains, where we can travel fast. We got away at 7.40 a.m., and found the country as described. This helped us along famously. The country is not unsuitable for railway construction; it is level and hard, subject to inundation, but not with rushing water, only back-water—still water, which can be easily dealt with. There are a good many creeks or inlets from the sea, in which the water is salt. These could be avoided by keeping the railway further inland, and so also could the ground which is subject to inundation be avoided. Of course, the land is poor, being spoiled by the occasional salt water; but there are some

ridges a few feet above the general run which are high and dry, and covered
with abundance of excellent herbage. We stopped for lunch rather early, and
rode across to look for the sea. It could be seen by climbing up a gumtree.
After lunch we drove closer to the sea, but could not get very near. We saw
what we thought was Point Parker, about fifteen miles off, but this is
uncertain. Our course afterwards was very erratic. We met with a great
many salt-water creeks, which we had to round or follow a long way up before
we could cross them. We came to a convenient camping-place soon after 3
p.m., and, as we were pretty tired, decided to camp for the night. Mr. Hann
went on again to look out for to-morrow. I cannot help fancying we are
within a few miles of our destination; to-morrow will surely show. We have
seen plenty of ironstone conglomerate ballast to-day, but very little suitable
timber—a few low "currejohn," bloodwood, &c. On the whole, I am agree-
ably surprised at the country. It has been extremely hot to-day. This is
Camp 60. Barometer, at 4 p.m., 905 ft., or 29.74 ; thermometer, 84°.

WEDNESDAY, MAY 4.—For several days past, the excitement and the
uncertainty have been rather too much, and is still growing. To find that
we must be near our journey's end, and not to know how near, is a little
too irritating. It seems to have reached a sort of culminating point this
morning. Last evening there were indications of rain, and as I had no
cover to my waggon I had to get a fly rigged over my mosquito netting. This
had a most depressing effect. I felt sure I should be smothered, so I had two
sides of it turned up; but in the night it came on to rain in earnest, and I
had to be closed in very much like a "black hole." How I passed the night
I don't know. I had the most horrible nightmare, for I had taken a lot of
strong cold tea before turning in. At daylight I was out as usual, and, mus-
tering all my strength and energy, got through my usual cold water sponging.
I did not, however, seem to rally after it; could not get warm, so went and
stood by the fire. I could not have been there many minutes when I fainted
away as dead as a stone, falling full into the hot ashes where the fire had
been for cooking the breakfast. Fortunately, Mr. Wyatt was close at hand
and pulled me out at once, and, after bathing my face and hands with cold
water, and giving me a little brandy, I was restored to consciousness, having
suffered nothing worse than two or three nasty burns on my arm and a
feeling of disgust at having broken down at the last moment. I presently
had another but milder faint, and then was pretty right for the day. I could
not, however, venture on horseback, and had to be very careful in the waggon.
The country was easy all day. At about 10.30 a.m. we got into a sand
hillock very near to the sea, and then, to our great delight, saw Allan
Island and Point Parker, and half way between them, at anchor, *The Pearl*,
about fifteen miles distant. This put new life into all, but we did not know
how far we had to travel. Along the beach it was clear enough, but we were
afraid to risk the tidal inlets we might have to encounter and perhaps to head.
Therefore, we kept inland and had to go many miles to head two or three. It
turned out afterwards that we might have gone along the beach quite easily.
However, at last we reached our goal, Point Parker, and after a little gesticulation, waving of all sorts of white things, &c., we had the satisfaction of
hearing two guns fired from the ship, and of seeing a flag hoisted in recognition,
whilst a boat put off for the shore with Captain Pennefather and others. Mr.
Hann and I soon found ourselves on board, as arrangements had to be made
for to-morrow. It was about 5 p.m. when we were first recognised by the
ship. I had got through the day wonderfully well. The excitement had
pulled me through, but now came the collapse. There was no mistake about
its being an attack of fever, brought on by fatigue and privation. It not only

attacked me, but Mr. Hann was very bad on board; and Mr. Wyatt and
Mr. Baynes, next day, when they came on board, were completely prostrated.
But I am mixing two days. During the 4th May we passed through pretty
level country, subject to being flooded to a depth of two or three feet, but not
rapid-running floods, rather back-water, flood-water ponded back by the tides,
not difficult to deal with in the construction of a railway : though the railway had better be kept back a few miles from the shore. Latterly we passed
a great many corkscrew palms or *Pandanus*. They bear a very nice-looking
fruit, not yet ripe. I am told the natives never touch it whilst they can get
anything else. We did not see a single native, but we were clearly close on
the tracks of some. In one instance they had their sticks and grass laid
ready for a fire, and we must have disturbed them. They get their water by
sinking a little depth into the sand a short distance from the shore. The beach
is stony, limestone conglomerate; plenty fit for ballast. Very little timber, some
of it bloodwood. The ground is a little bit boggy ; in fact, we got bogged,
and had to put in the "emergencies" only an hour before we reached our
destination.

THURSDAY, MAY 5.—I thought I might safely yesterday have telegraphed to
say we arrived in excellent health and spirits. I think I may still, but immediately after our arrival, even the very next day (*i.e.*, to-day), five out of
six of the white men of the party are down with fever — myself, Mr.
Hann, Mr. Baynes, Mr. Wyatt, and Davis, the cook. Mr. Hann pulled
through with his usual determination, and was well enough to go on shore
early with the captain, superintend the sending of the things to the ship, and
go for a ride. I was the next to struggle through, but there was little done
except to shake things down a bit. The cook is very bad. Mr. Baynes and
Mr. Wyatt are getting better; but the fever is a beastly thing. I have been looking at the captain's chart of the Harbour, and am told there is no mistake
about the accommodation for ships of any size. We hope to get the remainder
of the things on board to-morrow. The cook is to come with us, as he is not
able to accompany the party overland. I was really too ill to read the barometer on my arrival. I think, on the whole, our trip has been fairly successful.
We have not had any serious accident nor much sickness. The medicine
chest has scarcely been touched. A few doses of "Eno's Fruit Salt," or a
little aperient medicine, has sufficed. I attribute our success in a great
measure to the almost total abstinence from stimulants, as only in cases of
necessity was any touched, and about two bottles of brandy only were consumed by the party on the whole journey. We did not lose a single horse.
We left Mount Cornish with forty-five and brought every one of them to Point
Parker. One very important element was the purchase of four tried staunch
dray horses, which had been working together in the same team for several
months, and pulled together with marvellous unanimity. We called them the
"emergency team" because they were never yoked up unless we were in a
difficulty or had a bad river or creek to cross. Then they never failed. No
expedition should be without such a team. Several days must elapse before
we can reach the nearest telegraph station at Kimberly, the mouth of the
Norman River. I sincerely hope that, by that time, the fever will have left
us. I find I have been reduced in weight from 13 st. 7 lb. to 10 st. 12 lb.

1 8 8 1.

QUEENSLAND.

TRANSCONTINENTAL RAILWAY from ROMA to PT. PARKER.

(MR. WATSON'S REPORT ON TRIAL SURVEY OF PROPOSED ROUTE FOR)

Presented to both Houses of Parliament by Command.

Railway Survey Camp No. 7,
Charleville, 24th January, 1881.

PROGRESS REPORT No. 1.

ROMA TO CHARLEVILLE.

SIR,

I have the honour to report that I have driven over the country from Roma to Charleville, starting from Roma on the 14th January, and reaching Charleville on the 21st.

For the first sixty-seven miles (*i.e.*, seven miles west from Mitchell) the country is easy for the construction of a railway, there being, comparatively, no earthworks except surface forming and ditching; and this, indeed, may be said of nearly all the country from Roma to Charleville, with such exceptions as will be hereafter mentioned.

The first stream of any consequence is a creek called the "Bangowangeria," near Mount Abundance Home Station, and this is evidently flooded to a great height sometimes—I should say 35 ft. to 40 ft. above summer level; the next is the Maranoa River, near the township of Mitchell, which, although dry as a road for scores of miles in summer, still in floods overflows its banks and covers the country to a considerable depth for a long distance on either side.

There are several smaller watercourses, all of which show indications of great rushes of water at times.

I must leave for further information and consideration the question of dealing with the watercourses—whether to construct the bridges sufficiently high to provide for all floods, or to let the water in high floods go over the line, and submit to temporary delays and repair damages afterwards.

There is plenty of ballast all the way within easy distance, mostly basalt, but great scarcity of timber for bridges or sleepers.

After the first sixty-seven miles the country becomes scrubby and more uneven, poorer too, scarcely a blade of grass to be seen anywhere, but still plenty of stone ; and the line may be kept on, or nearly on, the surface, with a gradient of 1 in 50, with very little earthwork. There are a few ironbark trees, which could be used for sleepers, at many places along the line ; I think sufficient ironbark and box timber could be got abreast of the line to supply the necessary sleepers, but probably this would not include more than five (5) per cent. of the whole distance, the remainder would have to be provided from somewhere else, I am afraid, at considerable cost. There is very little country that can be called "ranges," very little that cannot be dealt with at comparatively small cost; the principal range is the "Angelalla Range," about thirty-five miles eastward from Charleville. The earthwork, if the line follow anything like the telegraph line and present road, will be considerable; but I am satisfied from what I have seen that an easier and more gradual ascent may be secured and a lower summit found for the railway. A detail survey of this portion of the line should be undertaken as soon as funds are available.

There is a great deal of country through which I passed similar to the Campaspe Plains, in Victoria, where excellent wheat is now grown, and which, as necessity may justify or require, will come under cultivation.

I am rather afraid of the ravages of the white ant, as in several of the townships the buildings are placed upon stumps or piles 3 ft. or 4 ft. above the surface of the ground, with projecting tin caps.

I noticed that from Black's Waterholes to Charleville a great many dams have been constructed by the Government with excellent results, and I cannot see why this cannot be done all over the country; all that is required is the outlay of more capital, and this, I fear, will never be attained until the holdings are very much smaller, and the security of tenure more satisfactory. To me it is evident that a man who has to make a living from 40,000 acres (*i.e.*, 12½ m. x 5 m.) must make it produce more per acre than he who has 4,000 square miles : this can only be done by extra expenditure, extra labour, and he, the former, has security of tenure whether he purchase or lease his land either from the syndicate or the Government, and every extra ton of produce goes to swell the revenue of the railway—a market must be found somewhere, and at the same time the general consumer is benefited.

My field notes and my previous experience lead me to the conclusion that the country from Roma to Charleville may be classed with the country from Stawell to Murtoa in Victoria; similar lines would cost about the same in the two localities if the price of labour were the same in both cases, but the the gauge of the line and the character of the permanent-way materials being different, allowance must be made.

I shall be able to prepare an approximate estimate of the cost per mile as soon as I can get some information which I have asked for from the Victorian Government.

On the whole, I believe a good, useful, efficient line can be made at reasonable cost ; that ⅓ (one-third) of the adjoining land is excellent, equal to any I have seen in the Australian colonies ; that ⅙ (one-sixth) is indifferent scrub ; and ½ (one-half) will eventually come under cultivation as it may be required.

I have, &c.,

ROBT. WATSON, C.E., M. Inst. C.E.

The Honorable The Colonial Secretary, Brisbane.

Camp No. 7, Charleville,
24th January, 1881.

GRAND TRUNK LINE TO ADELAIDE.

Copy of information obtained.

23rd January.—Saw Mr. McFarlane, who, in the Government service, surveyed the whole of the country down the Warrego River from Charleville as far as the boundary of Queensland, in the direction of Fort Bourke ; the boundary is at 29 degrees south latitude. He says the country is perfectly level all the way—at least the fall is perfectly uniform all the way ; there are few watercourses to cross, none of any consequence ; the line as sketched could easily be constructed ; there are a few inlets, or, as he calls them, billabongs, off the river, which are filled in time of floods, but the river scarcely ever exceeds its banks ; thinks it will be desirable to keep the line some distance from the river, and on the east side, until reaching Cunnamulla, then cross over in the direction of Eulo and Thorgomindah. He does not know the country beyond Cunnamulla, but as far as this it is mostly lightly-timbered flats or open plains, and what is known as "Mulga" country for some miles (seven or eight) on either side of the river. This is considered good country for cattle, and the fact of Mr. James Tyson having 79,000 head of cattle here in this locality leads to the conclusion that the country is good. He says there is stone not very far away, but he did not seem to know much about it.

24th January.—Called on Mr. Thornton, Inspector of Police in this place ; he had with him a Mr. Davis, who knows the country between this and Thorgomindah. Mr. Davis says the country is perfectly level from Charleville to Cunnamulla ; the east side is the best. The line would have to cross the Angelalla Creek near Mangalore ; no other creeks of consequence. There is no inundation of the flats until past Cunnamulla, further south towards Fort Bourke, then it breaks out over the flats ; there are small creeks, but of no consequence. The country from Charleville for the first fifty miles is sandy and poor, then from this to Cunnamulla it improves and gradually becomes rich black soil. There is timber—pines, bloodwood, etc.—all the way ; good-sized pines, large enough for sleepers, and bloodwood two or three feet through, all along the Warrego. No stone on the river, but stone a few miles back ; probably stone may be found along the line of railway—hard sort of stone and waterworn. The country a little way down becomes good for sheep —as good as any between Roma and Mitchell Downs.

From Cunnamulla to Thorgomindah, first to Eulo, the country is very level ; no watercourses of any consequence, they can scarcely be seen ; after crossing the Paroo River at Eulo, during or after a heavy flood, the country being low is inundated ; this continues for about four miles. There is a range between Bingara and Diniver Downs Stations, but there is a gap ; there are some insignificant creeks. From Diniver Downs to Thorgomindah there is nothing of any consequence to cross, neither ranges nor watercourses. There is no good timber all the way, indifferent box, bloodwood, and yapunyah timber : this is very hard and durable, plenty of it but not of large size ; it is something like white mallee, grows up to 15 in. in diameter. Plenty of stone, ironstone ; the soil is very good either for sheep or cattle, mostly devoted to cattle now. There is not at present any cultivation, except gardens taken care of at the stations by Chinamen—fine gardens, anything will grow ; directly a township is started a Chinaman's garden is started at once ; there is any quantity of vegetables grown, and the ground would grow anything if required ; another good timber is gidyah, very hard and durable.

Mr. Inspector Thornton was present the whole time, and from his own knowledge indorses what Mr. Davis has said.

P.S.—I have simply copied what I wrote down in my note-book ; I accept no responsibility ; I only say Mr. McFarlane and Mr. Davis appeared to be speaking the unbiassed truth. I may have more to say when I have seen the country ; meantime I hope this may be of service.

25-1-81.   ROBT. WATSON.

Mr. Davis suggests that it might be worth while to cross the Warrego above the Angelalla Creek and go down the west side.—R.W.

Railway Survey Camp, No. 17,
Blackall, 10th February, 1881.

PROGRESS REPORT No. 2.

CHARLEVILLE TO BLACKALL.

SIR,

I had the honour, on the 24th ultimo, of forwarding to you my first progress report, comprising the portion of the proposed Transcontinental Railway situated between Roma and Charleville.

On arrival at Charleville several of my horses were completely knocked up from scarcity of feed all the way, and several badly wanted shoeing. I was greatly disappointed to find there was no grass within a less distance than nine miles, and not a horseshoe nail in the township. I had to send my horses to the feeding ground and to procure nails from Roma; this delayed me until Thursday, 27th ultimo.

In the meantime I had learnt that there was great scarcity of feed between Charleville and Tambo, along the valley of the Ward River; I therefore determined to follow the Warrego and Nive Rivers.

The River Warrego, where it is crossed immediately after leaving Charleville, requires the same consideration as the other rivers, viz., as to crossing it at a high or low level; this must stand over for the present.

I followed the River Warrego, its western bank, for about twenty-six miles, then crossed it below the junction of the Nive River, and followed along, as near as I could, the eastern bank to Ellangowan, and thence to Burinda Station.

From Burinda to Tambo, viâ Caroline, on the Warrego, and the Nive Downs Station, I passed through some excellent land; the country is not difficult for the construction of a railway; there is abundance of stone to be procured without much difficulty, and I think sufficient timber for sleepers and fencing, if, indeed, the latter be considered necessary; but there are evidences of floods extending a considerable distance on both sides of the river, and I think the route I have followed deviates too much from a direct line from Charleville northwards to be suitable; I am inclined to recommend that it be not entertained.

There then remain two routes, viz.:—1st. The valley of the Ward River (western side), direct to Blackall, avoiding Tambo altogether; and, secondly, the dividing ridge between the Ward and the Nive waters. I have no doubt either of them would be preferable to one on the eastern side of the Warrego.

I have not seen the Ward Valley country, but everyone speaks highly of it; the course is very direct and the land excellent.

I have no doubt a line could be got along the dividing ridge between the Nive and Ward, and it would have this advantage, that very little provision would be required for the storm water; at the same time provision could easily be made for water along the line by reservoirs not far from it, or by pumping from the rivers where they are tolerably near at hand. All the land, or nearly all that I saw, was of excellent quality. I believe there is plenty of stone for ballast and sufficient timber for sleepers, etc. In saying there is sufficient timber for sleepers, etc., I think it may be worth while to consider the advisability of using the local timber as far as possible in the first instance and replacing it with better when, after the construction of the railway, carriage becomes more easy. The bridges should be so constructed that any piece showing symptoms of decay can be replaced without interfering with the traffic of the line.

Whilst on my way from Burinda to Tambo I had an opportunity of seeing two excellent reservoirs; everything shows that this is a step towards supplying the great want of the colony, and it will soon become an absolute necessity.

From Tambo towards Blackall the country improves in places; at Greendale Creek, about fifteen miles from Tambo, there was very fair feed; although only 10 (ten) days ago, before rain came, the country is said to have been as bare as any I have seen. There is every indication of abundance of grass generally at some seasons of the year, and indeed the country looks as if it only wanted a little encouragement to make it bring forth abundance of grass now.

From this to Blackall the country varies, sometimes level open plains, sometimes lightly timbered country, and now and then refreshing undulating downs, lightly, generally very lightly, timbered; and I have no doubt plenty of ballast will be found, for, in addition to limestone, there is a coarse sandstone and some waterworn gravel.

For a considerable distance we followed along the south bank of the Barcoo River, crossing it near the Northampton Station, and thence along the northern bank to Blackall; the country calls for no particular comment; the line should cross the river pretty high up.

The Barcoo, never, I think, rises to any very great height above its bed, probably not more than 15 ft. or 20 ft., but it overflows the country for a long distance, now on one side, now on the other.

On nearing Blackall the country for a considerable distance on the north side is at times heavily flooded; there is a series of anabranches running parallel with the river, extending for about a mile in width, the river itself being very insignificant.

I expect to leave Blackall about Monday next, and hope to reach Aramac by the end of the week. At Aramac it is my intention to leave the party for a while to explore for stone, etc., and to rest the horses, whilst I, in accordance with the expressed wishes of the Honourable the Premier, proceed by coach to Emerald to get a general idea of the country for a branch line; and I may possibly go by train as far as Rockhampton.

I think from Charleville to Blackall the country will admit of a railway being constructed at a slightly less cost per mile than it can be over the country from Roma to Charleville; and, on the average, the land, in my opinion, is nearly, if not quite, as valuable.

I have, &c.,

ROBT. WATSON, C.E., M. INST. C.E.

The Honorable the Colonial Secretary, Brisbane.

Withersfield,
5th March, 1881.

PROGRESS REPORT No. 3.

BRANCH FROM ARAMAC TO WITHERSFIELD.

SIR,

I have the honour to report that, in accordance with the wishes of the Honourable the Premier, conveyed to me before he left for England, on reaching Aramac and getting my party located where there was good feed for the horses, I proceeded by coach to Withersfield, the present terminus of the Central line, so as to be able to form an opinion of the country and its suitability for a railway.

I

For the first mile or two there are good open plains and rich soil, with plenty of stone for ballast and culverts, but after crossing a branch of the Pelican Creek what is called the "desert" commences, and extends for about 12 miles (twelve miles). It is very appropriately called the "desert"—a wretched barren tract of country, open sandy plains alternating with miserable scrubby gidyah trees of no value, larger than I have seen before, but rotten and useless ; not a blade of grass, and evidences of floods extending for miles, the *debris* lodged in the scrub 4 or 5 (four or five) feet above the surface of the ground.

There is very little difference in the level for the first 20 (twenty) miles ; there is then a little improvement in the quality of the soil, and a gradual rise for about 15 (fifteen) miles until the summit of what is called the "Jump-up Range" is reached, which I estimate, guided by the barometer, to be about 640 (six hundred and forty) ft. above Withersfield.

The formation for several miles, whilst crossing this range, is very peculiar, and I think well worth the attention of geologists. It has the appearance of having been at some time a large inland lake ; the basin, which the coach track traverses, is almost surrounded by bold sandstone cliffs, in some places 30 or 40 (thirty or forty) ft. in height, equal to supplying ballast and stone for culverts for a great many miles.

There is no steep pinch ; the greatest rise is about 170 (one hundred and seventy) ft. in about 2 (two) miles.

The highest point reached is about 35 (thirty-five) miles from Aramac, and up to this there is little or no timber of any value ; afterwards the country improves—then comes plenty of grass and plenty of ironbark timber suitable for every purpose.

For many miles the coach track passes through what is called spinifex country : there are various opinions as to the value of this grass; some say that it is valueless, others that it is good feed for horses and cattle. I noticed that the coach horses fed on it were all in excellent condition, and the soil looks fit, under cultivation, to grow anything.

About 15 (fifteen) miles after leaving the Jump-up Range, we crossed the Alice River, and this led to the following entry in my diary :—" The main transcontinental line from Blackall to Aramac must cross the Alice River; the road from Aramac to Withersfield crosses the Jump-up Range and afterwards the Alice River. It follows, then, that a line from Withersfield to the main line may be made without crossing either the Jump-up Range or the Alice River."

I afterwards learnt from Mr. Ballard, Chief Engineer for the Central line, that a line had been surveyed passing to the south of and avoiding the Jump-up Range, but crossing the Alice River.

After a conversation with Mr. Ballard, he says in a memo. : " Note for Mr. Watson—From this point 305 (three hundred and five) miles, the line could be slewed southwards, if necessary to effect junction with transcontinental south of the Alice and to avoid crossing that river." What I saw of Mr. Ballard and Mr. Hannam, engineer in charge of surveys, led me to believe that the Government may safely leave the matter in their hands.

After leaving the Alice River what is called the Main Dividing Range has to be crossed, but the approaches to it are so gradual that it is hardly noticed ; the barometer showed an elevation 100 (one hundred) feet lower than the Jump-up Range, viz., 540 (five hundred and forty) feet above Withersfield.

The easy way in which the Drummond Range is crossed by the coach track at Craven's Gap greatly surprised me ; in fact, I was over it before I knew anything about it, and I do not think the greatest elevation exceeded 340 (three hundred and forty) feet above Withersfield. With gradients of 1 in 50 (one in fifty) I do not think there would be anywhere more than 7 or 8 (seven or eight) feet of bank or cutting.

From the Drummond Range to Withersfield the country is tolerably easy, but there are some creeks which require serious consideration. There is basalt near Surbiton, and granite about 30 (thirty) miles before reaching Withersfield; there is also suitable timber—ironbark, bloodwood, &c.—nearly all the way. On the whole, the country is not difficult and the land is fairly good, with the exception of the 12 (twelve) miles of desert.

Mr. Ballard has kindly placed at my disposal plan and section of a line which he has surveyed from Withersfield to Barcaldine Downs; and, after this information, I have no hesitation in saying that, with the alteration avoiding the crossing of the Alice River, it is decidedly better than any line that could be got by following the coach track over which I have travelled.

As the coach would not return to Aramac before Thursday, the 3rd March, giving me 5 (five) clear days, I proceeded by train on the 26th ult. to Rockhampton. I was met and accompanied for many miles by Mr. Ballard and Mr. Hannam, from whom I obtained much valuable information, especially as to the provision for storm water and floods. The way in which the Nogoa and Comet Rivers have been dealt with leaves no doubt in my mind as to the desirability, as a rule, of keeping the line at a low level and letting the heavy floods pass over it.

A line similar to that from Rockhampton to Withersfield could leave nothing to be desired for the transcontinental traffic for many years; the speed and, as far as I could judge, the safety are sufficient. Beyond this I am scarcely justified in making any remarks; but the general construction, the dealing with the curves and gradients, show that the engineers knew well what they were about, and the whole effect cannot be otherwise than pleasing to the professional eye.

On my return to Withersfield on the 2nd instant, I found the coach could not travel in consequence of the heavy rain; and, as I may be delayed for several days, I have thought it advisable to write and forward this report.

The rain, which I find has been very general, although a temporary inconvenience, will no doubt help my progress after it is over, and from information I have received I could not have moved even if I had been with my party.

I enclose plan, which may probably afford some assistance, if my report be not sufficiently explicit.

I have, &c.,
ROBT. WATSON, C.E., M. INST. C.E.
The Honourable The Colonial Secretary, Brisbane.

Aramac,
18th March, 1881.

PROGRESS REPORT No. 4.

BLACKALL TO ARAMAC.

SIR,

It was my intention that my "Progress Report No. 3" should have comprised the country from "Blackall to Muttaburra," but circumstances have prevented it.

I made my intended inspection of the country between Aramac and Withersfield, and, whilst detained at the latter place under stress of weather, reported the result in "Progress Report No. 3," dated 5th inst.

My return from Withersfield to Aramac was not accomplished without some difficulties, in consequence chiefly of the dreadful state of the roads and temporary illness brought on by fatigue, and a complaint peculiar to the neighbourhood of the "Belyando River."

As I am delayed here for a day or two, I propose to forward a report of the country from "Blackall to Aramac," instead of to "Muttaburra," as I had originally intended. It will, I think, be short, for I am beginning to find a sameness in the country all the way—it is all suitable for railway construction; there is plenty of stone for ballast, and there is a scarcity of timber for sleepers and bridges.

We left our camp No. 18, near Blackall, on 15th February, and for many miles passed over open plains with tolerable feed and occasional waterholes; some watercourses, but none of any consequence—the principal one is about 15 (fifteen) miles from Blackall. After this we passed some very fine waterholes, in the neighbourhood of which the feed was of course all gone in consequence of travelling stock, otherwise there would evidently have been plenty of grass, because in one instance I saw a fenced-in paddock where there was plenty, whilst all around outside was perfectly bare. There are saltbush roots or stumps, but nothing left on them; there are some small trees and scrub, and both look healthy, and there is some timber that may be useful, but not much.

We passed over many miles of heavy sandy soil; but this is only bad for travelling over in a buggy; a railway can be made there well enough. We saw a good deal of "spinifex country." I cannot understand why this grass is not good. I have alluded to it in my report No. 3, when I had an opportunity of seeing more of it. There is, taking the whole distance from Blackall to Aramac, more useful timber than from Charleville to Blackall.

The only river of any consequence to be crossed is the Alice, near Barcaldine Station, and I can now, after seeing what has been done with the Nogoa and Comet, see what ought to be done with it.

There is plenty of stone; the quality of the land, I think, on the whole, is pretty much like that between Charleville and Blackall, and a line can be constructed at about the same cost per mile.

I followed the beaten track, which is very circuitous, bending very much to the west. On my way I noticed the same magnificent rolling downs extending as far as the eye could reach to the eastward, and there is no engineering reason why a line may not be made perfectly straight from Blackall to Aramac.

I feel it my duty to mention that I have been informed there is much better land further west, nearer to and on the Thompson River, but I have not seen it. A great deal of the land I have passed over is suitable for agriculture—that is, the land itself is suitable for growing anything; but it is possible that, although the fall of rain during the whole year may be sufficient, it may come at such uncertain intervals, and the droughts may be so long, as to render any attempts at cultivation quite out of the question.

I am certain that ample water can be stored at comparatively moderate cost for utilising the land profitably for pastoral purposes. The provision for agricultural purposes is a problem that I am not able to solve.

We had to camp 12 (twelve) miles before reaching Aramac, because there was no feed nearer to the township, and had to go 15 (fifteen) miles beyond it before we could find any more. It has been very disheartening; but now the rain has come this is all over, and I am afraid I may during the drought have been led to form a not sufficiently favourable opinion of the country.

The effect of the rain in the neighbourhood of Aramac has been truly marvellous—like magic; I could not have believed it. What three weeks ago was like a barren desert is now like a meadow.   I think I may hope to proceed without further trouble or delay.

I have, &c.,

ROBT. WATSON, C.E., M. Inst. C.E.

The Honourable The Colonial Secretary, Brisbane.

Camp No. 31, Winton,
31st March, 1881.

PROGRESS REPORT No. 5.

ARAMAC TO WINTON.

Sir,

I have the honour to report to you that, before leaving the main line for Withersfield, I drove over a considerable portion of the line from Aramac towards Muttaburra, and on my way afterwards from Aramac to Muttaburra I travelled the first part of the journey by coach; my buggy then met me and I drove over the remainder.

Whilst detained at Aramac I had an opportunity of examining some of the effects of the recent heavy rains; the Aramac Creek, quite dry when I crossed it some weeks before, had overflowed its banks and extended to a width of at least 2 (two) miles just below the township, although the greatest depth of water in the creek at the highest point of the flood could not have exceeded 15 or 18 (fifteen or eighteen) feet.

The country from Aramac to Muttaburra is really magnificent, with the exception of about 7 (seven) miles of gidyah and boree scrub, commencing 12 (twelve) miles from Aramac and continuing to 19 (nineteen) miles; but even in this scrub, wherever there is a small open space of only a few square yards, it is immediately covered with grass; there is also a good sprinkling of salt-bush.

At 20 (twenty) miles from Aramac the Mount Cornish Run begins, and I think it would be difficult to find a finer station property or one more perfectly managed. The soil is rich and feed abundant, and when nearing the station I saw hundreds—I think I might say thousands—of cattle grazing on the luxuriant pastures. Several dams have been constructed by the proprietors, and one large one by the Government near Sardine Creek.

There is also a Government dam near Aramac, which, I think, supplies the town with water. The reservoir is fed entirely by rain water, which should be pure, but after standing for some time it becomes quite hard and soap will scarcely lather in it. There must be something in the strata through which the hole was sunk from which the material for the dam was taken. I think this matter should receive consideration, because a great many of the railway water supplies will depend upon artificial reservoirs similar to this one. Some curious fossils are found here embedded in what I call limestone.

There is plenty of stone for ballast cropping up at irregular intervals, but there is no timber fit for railway purposes; a few posts might be got for fencing near the Thompson River.

There are only 3 (three) defined watercourses that can be called creeks, viz., one at Stainburn and the two branches of the Sardine Creek, within about 3 (three) miles of each other. There are many smaller watercourses on the Downs, which can be easily recognised and provided for.

The road appears to me to follow pretty high ground all the way. There is no reason why a line may not follow the Divide, or indeed any other route.

Between Muttaburra and Mount Cornish Station I had an opportunity of seeing something of what the recent flood was like. It appears to have joined the Landsborough and Thompson Rivers together, and extended for two or three miles in width. Very great provision will have to be made, if indeed the line is to cross anywhere near this place. It seems that after the Mount Cornish and Tower Hill Creeks join, the combination becomes the Thompson River; the Landsborough joins the Thompson some distance below Muttaburra. There is a difference of opinion about this; some say the junction is above Muttaburra. But there are two distinct rivers and a bridge over each on the road from Mount Cornish to Muttaburra. The crossing of these waters, if any be required, must of course be below the junction, wherever it may be.

We left Muttaburra on the 23rd March, and arrived here on the 31st March. For the first 20 (twenty) miles from Muttaburra towards Winton the country varies a good deal; at first it is slightly rangey, but the undulations are very gentle; there is some scrub and a lot of red conglomerate stone on the surface. The country becomes gradually more open, and finally develops into fine open plains.

For the next 40 (forty) miles towards Winton the country is difficult to deal with; fine open plains, splendid soil, a little sandy, level as far as the eye can judge; but there are evidences of the most extraordinary floods at all sorts of unexpected places. It is impossible to account for the direction the water takes. Where you would think it impossible any flood could reach, there are indications of a rapid stream having been running at considerable depth. I have met with somewhat similar country on a much smaller scale in Victoria, and experienced great difficulty in dealing with it. I believe if a careful section were taken along the road the most experienced engineer could not foretell from it where the water in time of flood would cross it, and where it would not; in many cases the water has evidently gone away through holes in the ground somewhat similar to what are called "crabholes" in Victoria.

I think it is a country to be avoided if a more suitable route can be selected, and I believe one can. I enclose a very rough bush tracing, which will enable you to trace two alternate routes on sheet No. 4 of new map of Queensland, either of which would, I think, be preferable to the one I am now travelling over.

No. 1 on the tracing is the one I am now on *via* Winton. The distance by it from A to B is 192 (one hundred and ninety-two) miles. No. 2, from A, *via* Muttaburra, S.W. of Mount Leichardt, crossing Mill's Creek near Manuka, thence across Wokingham Creek in a direct line to the point at which the Great Dividing Range intersects the 142° of east longitude, and joining line No. 1 at B on plan. Distance A to B, 184 (one hundred and eighty-four) miles. No. 3, from A going about due north crossing the river below the junction of the Towerhill and Mount Cornish Creeks, thence following the Divide between the Landsborough River and the Towerhill Creek until within about 60 (sixty) miles of Hughenden, then N.W. across the Landsborough River until the Dividing Range is reached, then following the Divide or near it until point B is reached. Distance A to B, 236 (two hundred and thirty-six) miles. (*These alternative lines are now shown on map of Queensland attached.*)

The provision for water would be very much less on No. 2 than on No. 1, and much less on No. 3 than on either of the others; but I am told the land is not nearly so good, and this is a matter that should receive the attention of the Government, as no doubt it will receive the attention of any syndicate that may undertake the work.

These distances 20 (twenty) miles and 10 (forty) miles bring me to the Divide between the waters of the Darr and Western Rivers—the waters of the Thompson and the Diamantina. Up to this point for a great many miles the feed has been very indifferent, in some cases almost superseded by weeds, yet the soil looks excellent. Immediately, or very soon after crossing the Divide, the Vindex Run is entered, and the scene changes at once ; the soil seems much the same, but the feed speedily becomes most abundant, and in one place for a considerable distance is literally up to the horses' girths. Everywhere, as far as the eye can reach on these open plains, there is as much feed as could be consumed without wasting a great deal by trampling it into the earth.

There is this difference between the two sides of the ridge ; on the western or Vindex side the watercourses are much more clearly defined, and you can tell where the water would cross the line ; on the eastern side you cannot.

This state of things continues until the station is nearly reached, about 15 (fifteen) miles from Winton ; here the ground on the south side of the Oondooroo Creek is subject to very heavy floods—over the top of the wire fences. On the north side of the creek the ground is much drier and firmer. We crossed the creek at the station. Afterwards we had to cross many tolerably well defined watercourses, and at Mill's Creek, a very short distance from Winton, we were completely stopped by a flood too dangerous for crossing.

From Aramac to this place I have scarcely noticed a single ant-hill ; I am therefore led to hope we may escape the devastations of this pest, the white-ant, if indeed we are to suffer from it anywhere. I am very much inclined to think we may not, and especially as far as sleepers are concerned.

Over the whole of the country from Aramac to Winton there is plenty of stone for ballast, but no timber, and the country is remarkably easy for the construction of a railway.

Some repairs have to be done to the harness, &c., which will occupy a few hours, then we proceed northwards, and, weather permitting, hope to reach Cloncurry within a fortnight ; my next report will comprise Winton to Cloncurry.

The party and horses are all in excellent condition.

I have, &c.,

ROBT. WATSON, C.E., M. INST. C.E.

The Honourable The Colonial Secretary, Brisbane.

Camp No. 43,
Cloncurry, 13th April, 1881.

PROGRESS REPORT No. 6.

WINTON TO CLONCURRY.

SIR,

I have the honour to report to you that I left Winton with my party on the 3rd April, at 4 p.m., and arrived at Cloncurry at 9.30 p.m. on the 12th April.

We followed the beaten track from Winton to Kynuna, and, crossing the Diamantina River near Dagworth Old Station, proceeded up the river on the west side as far as Belle Kate Station ; then crossed the river again and made for the " Great Dividing Range" in a northerly direction about 8 (eight) miles.

After crossing the divide there is no track—we had to travel by sun and compass, and our direct course for Cloncurry was nearly north-west, but, to avoid a range of mountainous hills which extends from the copper mine, curving to the east and north, we had to steer about north-north-west all the way from Fullerton Creek. We thus headed the range about 16 (sixteen) miles north-east from Cloncurry. We now took a south-westerly direction until we struck the road from Fort Constantine to Cloncurry. This we followed until we reached our present camp (No. 43) on an anabranch of the Cloncurry River, about 2 (two) miles from the township. We had been misinformed, or misjudged the distance by about 20 (twenty) miles, and had to travel the last 15 (fifteen) miles by moonlight.

The railway will have to head this range as we did, and I do not think can be taken nearer to Cloncurry than within about 15 (fifteen) miles.

The soil, with very few exceptions—one especially, when nearing Cloncurry —is really excellent ; the grass abundant, stone plentiful, and country easy ; but there is, as has been the case all the way, no timber suitable for bridges or sleepers.

We had considerable difficulties to contend with, the ground having not yet become sufficiently dry for comfortable and expeditious travelling ; the creeks in many places still flooded, and all the watercourses very boggy ; our harness constantly snapping, and no less than three poles broken between Winton and Cloncurry—indeed, but for the strong team of light draught horses purchased at Mount Cornish, we might at this moment have been struggling between Muttaburra and Winton ; they have already more than repaid their cost. The horses provided by the Police Department, with the exception of a little exuberant playfulness at starting, are excellent in their way, when the weather is fine and where the roads are good ; but in emergencies, such as we have frequently had to contend with, quite useless. The difficulty of getting any four of them to put their shoulders to the wheel all at once requires a deal of patience ; whereas, when the "emergency team" is once in harness, a gentle word and the slightest movement of the reins straightens the whole of the traces at the same moment, and the difficulty is over. No expedition of this kind should be without such a team. We sometimes had to put them in three and even four times a day. They are kept purely and entirely for emergencies, and thus are called the "emergency team." Of course, under ordinary circumstances, over fairly good roads, the police horses are infinitely to be preferred —they travel three times as fast and much more pleasantly.

The Werna and Wolkingham Creeks were both flooded ; but by great care and good luck we succeeded in crossing them without any accident. We were delayed for a few hours at Ayrshire Downs Station ; there is a smith's shop there, and some smith's work was badly wanted. The Werna and the Wokingham Creeks have both a lot of troublesome billabongs.

We crossed the Diamantina River because we were told that there was no road on the east side and there were four very difficult creeks to cross, whereas on the west side there were only two. About 20 (twenty) miles above the Kynuna Station we recrossed the Diamantina. At both crossings it consists of a lot of billabongs, so much alike that it is difficult to say when the real river is crossed.

All the country on each side of the river, and on each side of its tributaries near where they join the river, is evidently subject to terrible floods ; during the last heavy fall of rain there must have been many thousands of square miles submerged.

I have in a previous report (Progress Report No. 5) remarked on the effect of the floods on the country I have travelled over since I left Muttaburra, and suggested two alternative lines. The opinion I then formed has been confirmed and intensified at every step since—viz., the waters I crossed from Muttaburra to the crossing of the Diamantina must be crossed by a railway very much higher up, near their source, or headed altogether.

After crossing the Diamantina the second time we soon reached what has been called the "Great Dividing Range." This is a great mistake, and calculated to mislead; it might be called the "imperceptible ridge dividing the great waters" or the "Carpentaria divide." To the ordinary traveller—indeed, to the careful observer—unaided by instruments, it has no more visible existence on the ground than the parallels of latitude or the meridians of longitude which are shown on the maps. I do not think I should have noticed it if my attention had not been directed to it. Not so, however, with the surrounding scene. I do not know how to describe the view from this place; it certainly surpassed anything I have before seen in any of the Australian colonies; rich, undulating plains extending as far as the eye could reach to the east, north, and west, and as far to the south as where the timber fringes the plain near the banks of the Diamantina, richly clothed with the most glorious verdure, and just enough timber or bushes to give it the appearance of a magnificent park. We travelled in a north-westerly direction for about 20 (twenty) miles, and then seemed as far from the end of it as ever.

Of course I have seen this country under most favourable circumstances; what it may be in a season of drought I cannot say, but I have been informed that during the late dry season there was never any difficulty about feed; it was the want of water that was so distressing—there is, in fact, at present no permanent water. This will be cured by-and-bye; the surplus water will be stored, and what has to go away will be confined to legitimate channels, but this state of things will not come about until the holdings are different.

The next day we made for the Beaudesert Station, hoping to find there a place suitable for crossing the McKinlay Creek. We found a crossing, but a very difficult and rather dangerous one; but, with the kind assistance of the manager and some of the station hands, we were successful. The McKinlay, like all other creeks, has a lot of billabongs; in this case they extend about four miles, and in the recent floods there was one sheet of water for that distance.

The Gidyah Creek is also a series of billabongs—most troublesome ones. There is always in these cases considerable delay in searching for suitable crossing-places; many extra miles are travelled and much time is lost.

The country over which we travelled, as far as the eye could reach towards the north and towards the south, is all excellent and easy.

We had to cross Holy Joe's and Fullerton's Creeks, and the Williams and Elder Rivers—the last the most formidable of the lot; it looked for a long time impassable, but we succeeded in crossing it in safety. Each of these streams has its attendant billabongs, more or less obstructive.

The whole of the watercourses are tolerably well defined, and, consequently, comparatively easy to deal with. The grass is most abundant everywhere; in some cases horses which had no packs on them and no riders were completely hidden from a short distance by it, but I am afraid most of it is spinifex.

There is little variety in the timber between the M'Kinlay Creek and Cloncurry—occasionally coolibar, whitegum, box, and now and then a solitary bauhina, but nothing of any value except for fencing or firewood.

For many miles before reaching Cloncurry there is a great deal of spinifex country, and it does not look very prepossessing after the country I have been travelling over for the last few weeks; it looks dry and hard. About sixteen miles before reaching Cloncurry the country becomes a little more hilly and the soil poorer, covered on the surface with quartz and other pebbly waterworn stones. Near where we headed the mountain range the surface is covered with stones very obstructive to progress, and hurting the horses' unshod feet terribly. The stones are of various sizes, of a dark-brown colour, very hard and very heavy, containing iron or some metal.

Some repairs are required for the waggon and harness, but we hope to make a start early to-morrow morning.

I have said the line cannot be taken nearer than to within about fifteen

K

miles of the copper mines at Cloncurry, but there is no reason why a branch may not be made ; the country is admirably adapted for it.

All the party are in excellent health and spirits.

My next report will probably be written from Burketown, and include the country between this place and that, and we may anticipate reaching there within a fortnight of this time.

The mails are very uncertain and irregular ; our latest dates from "inside" are nearly six weeks old.

I have, &c.,

ROBT. WATSON, C.E., M. Inst. C.E.

The Honorable The Colonial Secretary, Brisbane.

Normanton, 13th May, 1881.

PROGRESS REPORT No. 7.

CLONCURRY TO POINT PARKER.

Sir,

I have the honour to inform you that we left Camp No. 43, near Cloncurry, on Good Friday, 15th April, and reached Camp 61, at Point Parker, on Wednesday, 4th May, at sundown.

It was my intention to have left Cloncurry a day sooner, but it was discovered at the last moment that some important parts of the ironwork of the waggon were damaged ; they had to be replaced here, as we should find no forge further on.

This delay gave me an opportunity of seeing and getting some information about the celebrated coppermine at Cloncurry. It is no doubt a very wonderful mine. Mr. Henry, one of the proprietors, was good enough to devote some considerable time to showing me a lot of the proceeds and explaining the state of affairs. Very erroneous and, I think, injurious reports have been in circulation, and, strange to say, believed, that the hill was one mass of pure copper, and on this account unworkable ; some enthusiastic scientific wiseacres have even gone so far as to write to the proprietors suggesting all sorts of cunning mechanical devices for hacking the monster to pieces. A visit to the locality, a not very difficult undertaking, would have prevented them from making fools of themselves and vexing the proprietors. That there is very rich ore there I know, because I have seen it, picked it up, have it now before me—ore that will yield probably 70, 80, 90 per cent. of copper. I have even seen pieces of pure copper, but only small pieces, and Mr. Henry has assured me that the richer the ore the more easy is the "getting" of it.

It is said the supply is inexhaustible and the demand insatiable. Of course I know nothing about this, but if the two be alike unlimited, then it cannot fail to be a source of considerable profit to a railway. At present the ore costs £8 10s. per ton carriage by dray to Normanton, 235 miles, and £1 10s. per ton in addition by sea to where there is fuel for smelting and a market for the copper. This, it is said, and I do not doubt it, leaves a very large profit. A railway could carry it to Point Parker, 310 miles, at very much less per ton. The mine has not been in active operation lately, in consequence, I am told, of one of the London proprietors having fallen into financial difficulties.

There is much other mineral wealth in the locality, but I think before coming to any definite conclusion the Government should satisfy itself as to the past results of "mining for metals other than gold" in different parts of the Australian colonies, and the causes of failure, if any.

After leaving our camp No. 43, we drove down on the eastern side of the Cloncurry over a capital level road for about sixteen miles, to a point near "Fort Constantine." Here we crossed the river, and here or near here I think the railway should also cross. The straight line which I have in previous reports alluded to as starting from Muttaburra to Aramac, would, if produced, cross near this place. The sandy bed of the river is about 300 yards in width ; the banks are not very precipitous, but the tableland is about 30 feet above the bed of the river, and this, it is said, has been covered to a depth of 2 feet or 3 feet. During the remainder of the day and during the following two days, steering nearly due west, we got along with varying success, and through country varying in richness and usefulness, but all suitable for easy railway construction. We passed some granite hills on our way rising suddenly above the open level plains. We crossed the Williams River, which is about half the size of the Cloncurry. There is a little bloodwood about here that might be made useful, especially on the western side of the river ; there is also a white gum in considerable quantities, of whose virtues or usefulness I know nothing.

The country is very uneven in many places between the Cloncurry and Dugald rivers, which latter we reached and crossed on the evening of Saturday, 16th April. The quality of the land appears to be fairly good all the way, but there is a disheartening proportion of spinifex grass.

Between the Dugald and the Leichardt there are insignificant ranges for several miles ; our course had a good deal of north in it. We passed over a lot of pebbly quartz spread thinly on the surface. The "several miles of rangy country" which we passed over is not at all difficult to deal with— in fact, the undulations are rather an advantage ; they indicate not only where provision for the storm-water *must be made*, but they also indicate unmistakably where provision *need not be made*. We crossed several creeks during the day ; one, the Cabbage-tree, rather a difficult one ; its name suggests that the cabbage-tree palm grows here, which indeed is the case, and very delicious it is.

We crossed the Leichardt River in the evening of the 17th April, and camped on its western bank, our object in crossing being twofold—to avoid some slightly rangy country, troublesome to drive over, which is to be found on the eastern side ; and further, to see something of what the western side is like.

It appears to me that all the country we have passed over, from McKinlay's Creek to the Leichardt River, is more suitable for cattle than sheep, but it is all suitable for a railway.

We continued to follow down the river on the western side for about fifteen miles, until we reached the junction of a considerable creek, and then recrossed. After this we continued for several days to pursue our course near the eastern bank all the way to Floraville. There is a large number of creeks joining the Leichardt River on its western side, some of them almost as large as the river itself, but few from the eastern side. There are some large tea-trees along the river in uncertain places. The timber looks good, but I could get no information as to its durability or usefulness, and I do not think any very large supply could be relied upon. The country all the way is very fine. The abominable grass seeds at this season of the year are almost unbearable ; they torture the cattle and horses almost to distraction, and I think would be fatal to sheep; they would creep into and through their bodies. But the land and its attendant advantages are very tempting ; the permanent water in the river, in splendid waterholes varying from two to fifteen miles in length, supplemented by numerous magnificent lagoons at various places along the tableland, sometimes miles back from the river, where the water lasts nearly all the year round, and which obviate the necessity for the cattle to go down and up the river banks ; add to this the abundance of good feed and the grateful shade of the trees, which are not inconveniently thick, what more can be desired for cattle country ? If I had to select 40,000 acres to live on with a five-mile frontage to the river and reasonable proximity to a railway, I would come out and look about here.

On the 21st April I took advantage of the kindness of Mr. Brodie, of Lorraine Station, to communicate our progress to you by telegram, *viâ* Normanton. At that time it appeared quite certain that we should reach Floraville on the 23rd, and we thought Point Parker on the 28th April at latest ; but heavy rain came on in the night of the 21st and threw out all our calculations most completely. We lost at least four days through a fresh flood in the Leichardt, and the boggy, clogging state of the rich brown soil caused by the heavy fall of rain. I need not encumber this report with details of our difficulties either here or in crossing the Alexandra or Landsborough, the Leichardt and the Albert Rivers. We crossed the Alexandra without any difficulty just above its pretty fall. We nearly got washed away in crossing the Leichardt, about four miles below the falls ; another two inches would have "done for" us if we had been foolish enough to attempt to cross, which I think we should not. At the Albert river the great scare was the reported danger from alligators, which was exceedingly childish ; they do not like any commotion ; their habits are retiring ; they like to sneak their prey, and to enjoy their food in ease and comfort. We crossed both, however, in safety without any serious accident.

About twenty miles above the falls on the Leichardt River there are some bold perpendicular rocks or cliffs on either side of the stream, and a narrow ravine, say sixty yards wide, between them, through which ordinary floods pass. This is the most suitable place I have yet seen for a railway bridge, if one be required ; but whilst there I was led to make inquiry about the country between this and the Gulf of Carpentaria at a place called on the map "Morning Inlet." I thought it possible that the cost of the three bridges —namely, over the Leichardt, the Albert, and the Nicholson rivers, say altogether £150,000—might go far to bring Morning Inlet to something like what Point Parker is in its natural condition. The information I obtained was not encouraging.

The country between Floraville and Burketwn is rich enough, but very difficult to travel over in any wheeled conveyance ; it is rough enough to shake all the amiability out of any ordinary individual. It is called locally the "Devil-devil" country, and it becomes the name. It is not desirable that a railway should be made over it ; it is all said to be subject to heavy floods. There is more suitable ground about twenty miles further south, abreast of the crossing place I have spoken of on the Leichardt River.

We reached Burketown late in the evening of the 29th April. There is not the least necessity or justification for bringing a railway near it. The Albert River should be crossed about sixteen miles above it, near the junction of the Brook and Barclay Rivers—which junction indeed forms the Albert River. We could get little or no information at this place about the country between it and Point Parker ; all seemed dark. A Mr. Harris, who, it is said, could have given us much valuable information about the route, was unfortunately crushed to death underneath waggon wheels only two days before we met his mate at the Leichardt crossing.

The next day Mr. Hann and I rode for several hours, all day in fact, examining, chiefly, points for crossing the Albert and Nicholson rivers.

It appears to me that one straight line can be run from the crossing-place about twenty miles above the Floraville Falls on the Leichardt, crossing the Albert say two or three miles below the junction of the Brook and Barclay rivers, to about twenty-four miles after crossing the Nicholson River, the general bearing being about N.N.W., from thence another straight line bearing nearly due north to Point Parker.

The country between the Albert River and Point Parker is not by any means unsuitable for railway construction. The whole distance would be less than seventy-five miles, and the first forty miles are very easy, with plenty of ballast—ironstone conglomerate—all the way. Afterwards, the country over which I travelled near the Gulf lies low and is subject to floods, but "still-water floods," tidal waters assisted by the storm waters—no rapid, destructive torrents ; in fact, I think the earthworks for a line might be constructed by an embankment 4ft. 6in. high, made from side-cutting, at a cost of about £500 per mile ; no doubt it would be cheaper and better a mile or two farther back. Ballast can be found anywhere.

I have no means of making any estimate of the cost of the bridges or other works at hand, as I did not find a single letter here ; this may probably afford material for a further and final report.

Our expedition has been remarkably free from any unpleasant incidents ; only one accident worth mentioning—the groom's horse fell with and rolled over him, on the 18th April ; but nothing was broken, and with a few days' rest he is now all right again. We have never touched the medicine chest, except for a few pills and a little goldbeater's skin. Drove four-in-hand in the same inexpensive express waggon all the way from Roma to Point Parker, proving that there is no Mount Cenis all the way. The Queensland Government schooner "Pearl" awaited our arrival, and Captain Pennefather was all kindness and hospitality.

Nothing could exceed the everlasting energy of Mr. Frank Hann, or his indomitable will and perseverance. He seemed to have but two thoughts on his mind—one the progress and success of the expedition, the other my personal safety and comfort. This spirit was readily, eagerly caught up by every member of the party. My thanks are due to all, and especially to Mr. Baynes and Mr. Wyatt, who, through the kindness of the Honourable the Premier, came with the party, and have done good and willing service.

Captain Pennefather has shown me his chart. It looks very satisfactory. I had an opportunity of verifying some of his soundings, and have every confidence in them. I must, however, leave it to others to decide on the merits of the "harbour." The accuracy of the information submitted I should entirely rely upon.

I hope to reach Brisbane by the steamer "Almora" on the 8th proximo.

I have, &c.,

ROBERT WATSON, C.E., M. Inst. C.E.

The Honourable The Colonial Secretary, Brisbane.

QUEENSLAND.

TRANSCONTINENTAL RAILWAY from ROMA to PT. PARKER.

(MR. WATSON'S REPORT ON TRIAL SURVEY OF PROPOSED ROUTE FOR)

Presented to both Houses of Parliament by Command.

SUMMARY.

Brisbane, 28th June, 1881.

SIR,

From time to time during my travel across the continent from Roma to Point Parker, I have had the honor of forwarding to you progress reports—in all, seven—describing the general features of the country through which I was passing. It may now be convenient if I write a short summary of the whole, describing the impression left upon my mind.

I left Roma on the 14th of January last, and arrived at Point Parker, on the Gulf of Carpentaria, on 4th May. My party on starting consisted of Mr. Baynes and Mr. Wyatt (two young gentlemen who, with the kind permission of the Honourable the Premier, accompanied the party to give all the assistance in their power), three white men (a cook, groom, and coachman), and a blackboy. On our way we had the assistance of the local police; and on our arrival at Blackall, on 16th February, we were joined by Mr. Frank Hann and a blackboy. Mr. Hann is one of the most accomplished bushmen in the colonies; and from this time throughout he acted as guide and took general charge of the men and horses, at the same time giving me all the assistance which his local knowledge enabled him to do.

On leaving Roma I proceeded by the coach road and telegraph line to Charleville. For the first seventy miles the soil is as rich as any to be found in any of the Australian colonies, very lightly timbered, and all very easy for railway construction. There appears to be plenty of stone suitable for ballast, but a scarcity of really good timber for bridges and sleepers. I may here remark that this want exists to a great extent throughout the whole distance from Roma to Point Parker. The difficulty may be met to some extent by using the local timber, a practice that has been followed with advantage for several years in Victoria. After passing through the Mitchell downs, or about seventy miles from Roma, the country becomes scrubby and slightly rangey, the soil less rich. This continues off and on for several miles; but occasionally, frequently, there occur patches, several miles in extent, of land suitable for cultivation—in fact, very similar to the Campaspe Plains in Victoria, where excellent wheat is now grown.

About thirty-five miles before reaching Charleville we reached the Angel-alla Range, which is the only range of any consequence met with on the way to the Gulf. If the line follow the direction of the present road there will, with a gradient of 1 in 50, be some heavy earthworks; but I am confident that a careful survey will show that a lower summit, with a better approach to it, can be found, thus materially decreasing the cost of the work.

Where I have spoken of "really good timber for bridges and sleepers," I have had in my mind ironbark, cypress pine, bloodwood, ironwood, redgum, &c. These, I fear, could only be procured at considerable cost. In the absence of these and of any local timber, it may be worth while to consider the propriety of using cast or wrought iron sleepers. They are only about one-third of the weight of ironbark, and the cost of carriage is a serious item when it extends over hundreds of miles.

From Roma to Charleville there had been no rain for several months, and there was consequently a scarcity of water and feed for the horses; but this did not alter the quality of the soil.

Before leaving Charleville for Blackall, I learned that the valley of the Ward River was the most suitable for a railway, that the soil was much richer than by any other route, but that it was just then impossible to travel that way in consequence of the scarcity of feed and water. I therefore determined to follow the Warrego and Nive Rivers, and proceed by way of Ellangowan, Burenda, Caroline, Nive Downs, and Tambo, to Blackall. I found the country excellent all the way, for some distance lightly timbered, then magnificent open rich rolling downs; and I think a sufficiency of water might be stored at a trifling cost during the wet season to supply the wants during drought. There is sufficient local timber within reasonable distance for sleepers and fencing nearly all the way. The country is easy for the construction of a railway, the earthworks light, and indications of plenty of ballast. The route I followed is, however, far too roundabout for the railway. There are two others, either of which would be preferable, viz.—one following the divide between the Ward and Nive Rivers to Tambo; the other through the rich land on the west side of the Ward River straight to Blackall. I prefer the latter, but the line should be kept well back from the river: in fact, as a rule, the "divides" between rivers should be followed wherever practicable. This would, of course, present a difficulty, and necessitate artificial means for providing water, such as tanks, dams, or wells.

At Blackall I was joined by Mr. Frank Hann, who relieved me from the general management of the party, and was of immense assistance. The country from Blackall to Aramac varies a good deal in places, but on the whole is good for pastoral purposes, or, indeed, for agriculture when a market can be found for the produce, and the means of conveying it are provided. The scarcity of feed still continues, and can scarcely be expected to improve till the rain comes. I see no reason why a very nearly straight line may not be followed all the way from Charleville to Aramac, passing within a few miles of Blackall.

On reaching Aramac, as there was no feed in the locality, the party proceeded to Mount Cornish, where accommodation was offered for the horses; and I, as requested by the Honourable the Premier, went to Withersfield, to be able to form an opinion as to the most suitable route for an extension of the Rockhampton line from Withersfield to a junction with the transcontinental line. This is not properly a part of the main line—particulars respecting it will be found in my Progress Report No. 3.

Whilst I was away from my party the welcome rain came, and I was detained for some time at Withersfield and Aramac. I reached Mount Cornish on the 20th March, and as the rain appeared to be over, and the ground fast drying up, we started on the 23rd March on our journey.

The country from Aramac to Mount Cornish is exceedingly rich all the

way, and mostly consists of magnificent open downs fit for any purpose, either
pastoral or agricultural. About twelve miles from Aramac there are a few miles
of scrubby country, but the clearing would not be costly, and the soil is rich,
as is evident from the fact that wherever there is a small space free from
scrub it is covered with luxuriant grass.

It had been my intention to proceed by way of the "divide" between the
Tower Hill Creek and the Landsborough River, both tributaries of the Thomson River ; but from information I received, I decided to go by way of Winton
and the Diamantina River.

The journey from Muttaburra to Winton was a very difficult one: it rained
a great part of the way, and the rich brown soil clogged the wheels of the
waggon, making it exceedingly difficult to make any progress. The soil all
the way is all that could be desired, and the country slightly undulating, very
easy for railway construction; but there are indications of floods in extraordinary places, which, added to my subsequent observations of the Diamantina and its branches, lead me to the conclusion that it would be better to
cross all these waters higher up, nearer their source, or to head them altogether. Indeed, it appears to me that, to accomplish this, a straight line
might be very nearly followed all the way from Aramac to Fort Constantine,
on the Cloncurry River, and still continued until the "divide" between the
Cloncurry and Leichardt River is reached; this would miss the Diamantina River altogether, and cross some of its principal tributaries, viz.—the
Wokingham, Werna, and Mills' Creeks, near their sources; the country
through which such a line would pass is very suitable for railway construction,
and the soil exceedingly rich—in fact, similar to the Diamantina country
over which I have passed—embracing a portion of the Ayrshire Downs, whose
richness is well known; but there is still a scarcity of good timber.

After crossing the Great Dividing Range, which is in reality imperceptible,
I followed a nearly north-westerly direction, passing through excellent country
all the way until I reached the Cloncurry River, which I crossed near Fort
Constantine, leaving the Cloncurry copper-mines about sixteen miles to the
south, and still pursued the same general direction until I reached the
Leichardt River, which I crossed and followed down the western side for
about twelve miles, when I came to a large creek, and, having learnt that
there were many others nearly as large as the river itself running in from the
western side, I recrossed the river and followed down the eastern side until
I reached Floraville or Chandos, when I again crossed about four miles below
the falls near the junction of the Alexandra or Landsborough River.

The country all the way from the Great Dividing Range, which, for dis-
tinction, I must still call it—very much surprised me ; I was, somehow, under
the impression that the soil was of indifferent quality, and the country rangey
—in fact, mountainous ; the very opposite is the case. As far as the
McKinlay Creek the country is admirably adapted for pastoral purposes,
either for cattle or sheep, and for cultivation also ; it is almost level, and very
easy for the construction of a railway, with every appearance of plenty of
stone. After passing the McKinlay Creek, all the way to Chandos, I think
the country is more suitable for cattle than for sheep, because there are in
many places grass-seeds which, I think, would eat into and kill sheep ; the
pasturage is more suitable for cattle ; sheep would be lost in it. There is
also a considerable extent of spinifex country; but for cattle I fancy it could
scarcely be surpassed, and especially on the eastern side of the Leichardt
River, where permanent water in magnificent waterholes, supplemented by
numerous lagoons, in many cases, several miles back from the river, supplying
water nearly all the year round, and the rich soil and luxuriant pasturage,
appear to me to leave nothing to be desired.

About eighteen miles above the Falls, which are at Floraville, there is a suitable place for a bridge—the most suitable I have yet seen—bold rocky cliffs about sixty yards apart, with a ravine between them which ordinary floods pass ; and if the river be crossed here—I say *if*, because it has been suggested that the line might be taken straight into the Gulf, to a point called Morning Inlet, but the result of inquiries I made was not encouraging, and I concluded that the expenditure of a sum of money equal to the cost of crossing the Leichardt, the Albert, and the Nicholson Rivers would not place Morning Inlet as a port in a position equal to that in which Nature's laws have placed Point Parker ;—if, therefore, the river be crossed here, a straight line may, I think, be taken all the way in a north-west direction to within about fifty miles due south from Point Parker, crossing the Albert near the junction of the Brook and Barkly Rivers, and the Nicholson immediately above the effluence of a large stream, which soon becomes as large as the Nicholson itself, and finds its own way into the sea — by some it has been called the Western Nicholson—leaving Burketown about fourteen miles to the north. The country from Floraville to Point Parker varies a good deal ; from the Leichardt to within about forty miles of Point Parker I should say it is good cattle country. A considerable portion of it is subject to floods, but for a great many miles after crossing the Nicholson it is undulating, high and dry above all floods. It has been said that the country from Burketown to Point Parker is impracticable for the construction of a railway, in consequence of the floods. I consider this a great misrepresentation, entirely unfounded. There are, no doubt, portions, several miles, which are subject to floods, but not raging torrents, carrying everything before them—rather, still water, tidal floods, caused by the storm waters being met by the tidal waters, not presenting any insurmountable or costly obstacle to the construction of a railway. The route I followed for the last thirty miles was near the Gulf, and we had to contend with several saltwater inlets. These can be avoided by keeping a few miles further to the westward. The last forty miles of country is probably not quite so suitable for cattle as what has preceded it—there is a good deal of brackish water—but it may be good for other purposes.

The country immediately around Point Parker is not low, or, I think, ever subject to floods. I have no doubt a suitable site could be found for a township.

Of the harbour itself I scarcely feel called upon to express an opinion. I am told by experts that the soundings on Captain Pennefather's chart indicate sufficient water for every requirement. I had an opportunity of verifying some of these, and should be inclined to place the utmost reliance on their accuracy.

I have so far omitted any mention of the rivers which will probably have to be crossed between Roma and Point Parker. The principal are the Maranoa, the Warrego, and the Ward (unless it be decided to cross below their junction), the Barcoo, the Alice, the Thompson, the Cloncurry, the Leichardt, the Albert, and the Nicholson. There are a great many creeks which almost deserve the name of rivers. I can scarcely venture, in the absence of any survey or sections, to estimate even approximately the extent or cost of the bridges. The floods are at times very high, and such floods have in other cases been very successfully and economically dealt with by constructing the bridges sufficiently high for ordinary floods, and letting high floods pass over the line, submitting to temporary inconvenience, and repairing damage, if any, as quickly as possible afterwards. I think, in most cases, the bridges, both for rivers and large creeks, should be of timber ; but the culverts may be economically built with stone abutments, and tops either girders made out of two wrought-iron sleepers back to back, where iron sleepers are used, or three

L

ordinary rails bolted together, when the span does not exceed six or seven feet ; but this is matter of detail.

The Cloncurry mines demand some notice. I had an opportunity of going over the principal one, and, through the courtesy and kindness of the managing proprietor, Mr. Henry, of gaining some valuable information. It is, no doubt, a very rich mine ; but its richness has been greatly exaggerated, to the annoyance of the proprietors. It has been said, and—strange to say—by some believed, that it was one mass of pure copper, which it was impossible to work. A moment's consideration will, I think, show the absurdity of this. That there is very rich ore there I know, because I have seen it—ore that will, I am told and believe, yield 70, 80, 90 per cent. ; I have even seen some specimens of pure copper, but only small ones. Mr. Henry informed me that the ore from this mine would yield probably 30 per cent. of copper. I am no judge. It is easily got (the richer the ore, the more easy the "getting") ; the carriage by dray to Normanton is about £8 10s. per ton, and thence to Sydney about £1 10s. per ton. I fancy this must leave a large profit ; and if the demand and supply be as unlimited as it is said they are, then the carriage to a seaport must necessarily bring a large revenue to a railway company, even from this one mine ; but it is said, and so universally and persistently said as to almost compel belief, that the country is full of such mines, and also many of iron and other minerals. The route I have indicated on the map takes the line not nearer than within sixteen miles of the mine. It is probable that a more comprehensive survey may show that it is practicable to take the line nearer to or even quite through the mining district, and further information may justify it. But so many mining speculations have been started in the colonies and failed, that I think, before deciding to deviate from a direct course for the main line, inquiry should be made as to the results of "mining for minerals other than gold," and the causes of success or failure. In answer to my question why such a rich mine had remained for so many years comparatively inactive, I was told it was in consequence of the failure of one of the London proprietors, but that there was every prospect of its being in full working order again shortly ; and no doubt the prospect of a railway and cheaper carriage will help to encourage this. The country is admirably adapted for the construction of a branch line.

I have suggested that the work of first construction might be facilitated by the use of the local timbers, but I have not lost sight of the importance of securing really good timber, even at considerably increased cost—timber such as ironbark, cypress pine, bloodwood, &c. ; or of the use of iron for sleepers and culverts.

On my arrival at Point Parker, as I had several days to spare before it was necessary I should be at Thursday Island to join the mail steamer, and the Q.G. schooner " Pearl " was at my disposal, I determined to try if I could find any suitable timber on the shores of the Gulf.

At the suggestion of Captain Pennefather I determined to visit the Batavia River, whose mouth is on the eastern side of the Gulf, about eighty miles south from Thursday Island, and four hundred miles from Point Parker.

We first visited the Norman River, which we had difficulty in reaching, as we chanced to reach the bar at low water, and had to anchor there for about twelve hours. On reaching Kimberley we left the ship, and in the whaleboat went up the river to Normanton. It is a fine river, but the bar is a great obstacle, and I did not find any timber, which was what I was in search of. On leaving Kimberley we were again delayed for about twenty hours in consequence of there not being sufficient water to let us cross the bar, there being only one tide in the twenty-four hours.

On nearing the mouth of the Batavia River we were met by a strong head wind. We had to beat all the way for several miles, and as the soundings were taken every few minutes we got a series of cross-sections of the channel.

The *modus operandi* was to go in one direction until we reached three fathoms water, then to tack and proceed in a nearly opposite direction until we reached a three-fathom point on the other side of the channel, then tack again, and so on. In crossing from one three-fathom point to the other we invariably crossed in mid-channel a depth of five, six, seven fathoms, and, as we neared the entrance, nine and ten fathoms. The mouth of the river is about two miles wide, and immediately inside is a large basin, oval in shape, about seven miles by five miles, with a very large extent of deep water.

After crossing this basin, with five or six fathoms of water all the way, we proceeded about twelve miles up the river, anchored, and went on shore. Here, on the south side of the river, we found a fine forest of timber—bloodwood, ironwood, and stringybark or messmate—I think, sufficient to supply as much timber for sleepers, &c., as it will be found desirable to carry from the Point Parker end of the line.

The importance of using timber that will resist the ravages of the white ant and dry-rot can scarcely be over-estimated. The use of iron sleepers is also deserving of consideration. The price per sleeper for iron at Point Parker would, no doubt, be considerably greater than the price per sleeper for timber, but they are much lighter, and the carriage is a heavy item; however, given the cost of each description at Point Parker, the weight of each, and the cost of haulage per ton per mile, and it is easy to determine beyond what mileage on the line it would be more economical to use iron than timber.

The cost of water carriage from Batavia River to Point Parker (four hundred miles) would be trifling.

We proceeded about forty miles further up the river in the boat. There is plenty of water all the way, three fathoms where we turned back; and the banks, where the timber is found, are very convenient for shipping. I think a further survey of the river should be made; it is by far the finest I have seen in the colonies.

We did not see many blacks up the river, but those we did see were fine, muscular fellows, with limbs such as I have not seen before; self-reliant looking. They showed every disposition to be friendly, and if treated fairly and kindly would probably be useful allies, but if treated harshly would be very troublesome.

Fuller particulars will be found in a separate report furnished by me on the 15th instant.

In conclusion, it appears to me that the whole question may be dealt with in two paragraphs:

1st. The general features of the country.

2ndly. The facilities for railway construction.

What I have wished to convey is:—

First.—The country is comparatively level from end to end, there being only one range worth mention, viz., the Angelalla Range, about thirty-five miles from Charleville. The soil is, for nearly the whole distance, surpassingly rich and suitable for either pastoral or agricultural purposes; the southern portions for either cattle, sheep, or the cultivation of ordinary crops: the northern for cattle or tropical products, sugar, cotton, rice, &c., &c. As a proof of the richness of the soil and the value of the land, may be cited the avidity with which all station properties have recently been purchased by capitalists, immediately on being offered for sale: and this with the uncertainty of a holding under a squatter's license or lease. When the holding is made secure, either by purchase of the freehold or a lease giving security from invasion or interference for ever or for a definite number of years, the expenditure of capital and labour will be justified, and the productiveness of the soil unlimited. Perhaps I saw a great part of the country, shortly after a fall of rain, under somewhat favourable circumstances; and my only fear is that, in my anxiety not to exaggerate anything or to present an overdrawn picture, I may have erred in the opposite direction, and understated the richness of the country over which I have travelled and through which I think the line can be taken.

Secondly, as to facilities for construction. It will be gathered that I consider the country all the way remarkably suitable for the cheap construction of railways. It is so level or so uniform, easy in its undulations and in crossing the divides between waters, that the earthworks must be comparatively light—surface-forming nearly all the way. There are indications of plenty of stone for ballast and the abutments of culverts; the quantity of ballast now being used in the more recently constructed lines scarcely exceeds 500 cubic yards (five hundred cubic yards) per mile. The principal, perhaps all, the bridges can be constructed of timber in such a way that the high floods may pass over them. For many miles, if the local timbers be adopted, sufficient will be found within reasonable proximity to the line; and, if more durable timber be required, plenty can be found at the Batavia River, involving cheap water carriage (about four hundred miles) at the Gulf end; and in the neighbourhood of Roma at the sourthern end; of course, the carriage by railway would be costly, and it may be worth while to consider the propriety of using wrought-iron for sleepers and for the tops of culverts. The line may be very direct—most of the rivers admit of being crossed at any one point as well as at any other; a certain sectional area of water-way has to be provided, and it does not matter much exactly where. I think the general direction may be in straight lines, or nearly so, as follows :—

(1.) From Roma to the divide between the Ward River and western waters, nearly due west, 168 miles.

(2.) From this point up the valley of the Ward, and crossing the Barcoo and Alice Rivers straight to Aramac, bearing a little west of north 232 miles.

(3.) From Aramac to the divide between the Cloncurry and Leichardt Rivers, crossing the Thomson and Cloncurry Rivers, bearing W. 30° N. about 370 miles.

(4.) Following the divide between the Cloncurry and Leichardt to within about 60 miles of Chandos, nearly due north 91 miles.

(5.) Thence to a point nearly due south about 50 miles from Point Parker, crossing the Leichardt, the Albert, and the Nicholson Rivers, bearing about north-west 104 miles.

(6.) From the last-mentioned to Point Parker, due north, about 50 miles.

These distances are scaled from the map, and are only approximate.

I attach a plan showing roughly the route I followed, in firm red lines, and that which I have suggested, in dotted red lines; and with this you will receive a copy of my rough field-notes taken daily during the journey. I prefer to send them to you just as they were jotted down, without any alterations, with the exception of striking out some irrelevant paragraphs.

I have, &c.,

ROBT. WATSON, C.E., M. Inst. C.E.

The Honourable The Colonial Secretary, Brisbane.

1 8 8 1.

QUEENSLAND.

TRANSCONTINENTAL RAILWAY from ROMA to PT. PARKER.

(MR. WATSON'S REPORT ON THE SUITABILITY OF TIMBER ON THE
BATAVIA RIVER FOR THE CONSTRUCTION OF)

Presented to both Houses of Parliament by Command.

TRANSCONTINENTAL RAILWAY.

Brisbane, 15th June, 1881.

Sir,

You will have gathered from the progress reports which from time to time
I have had the honour of submitting for your information, that the great
difficulty attending the construction of a transcontinental railway from Roma
to Point Parker is the almost entire absence of timber suitable for sleepers or
bridges, culverts, &c.; in fact, it may be said that practically there is none all
the way. It will have to be carried from the two ends, unless the railway be
extended from Withersfield to a point at which it would cut the Transcontinental line probably somewhere near Barcaldine Station, in which case
two fresh points would be available for carrying on the construction, viz., one
in each direction.

The subject of sleepers, a heavy item, has received some consideration. I
have been divided in opinion between the use of timber and iron. The cost of
carriage must determine this.

The cost of iron delivered at Point Parker or at Roma would no doubt be
greater per sleeper than the cost of timber, but the iron sleepers are very
much lighter than timber, and there must be a certain mileage at which the
first cost of the timber sleeper, with the cost of its carriage added, would equal
the first cost of the iron sleeper with its carriage added, after which point it
would clearly be more economical to use the iron sleeper. I am assuming so
far that the two are equally durable. This is, of course, open to correction—
there may be some advantages which one has over the other ; no doubt the
iron sleeper is proof against the ravages of the white ant, but it is said that
certain colonial timbers are also free from the attacks of these pests. However,
returning to the subject of the scarcity of suitable timber along the proposed
route, I determined, as I had several days to spare, and the Queensland
Government schooner " Pearl " was, through the kindness of Captain Pennefather, placed at my disposal, to try if I could find any suitable timber on the
Gulf of Carpentaria.

At the suggestion of Captain Pennefather, I determined to visit the Batavia
River, which is within less than 100 (one hundred) miles of the northern

extremity of the Peninsula, and I think about 300 (three hundred) miles from Point Parker; I was rewarded by finding, along the south bank of the river, a forest that I believe will supply all the timber that will be required for the line, or that it will be desirable to take from the Point Parker end. There is a variety of timber, but it is chiefly bloodwood, ironwood, and messmate or stringybark; it is all easily got at from the river, and I can see no reason why it should not be delivered at Point Parker at as low a rate as the average price of hardwood in any other parts of the colony. The cost of extra carriage on the railway is a matter for additional consideration. The river itself, as a means of bringing the timber away, demands some remarks; it is the finest I have seen in the colonies, but this is not saying much, because I have not seen many.

There is practically no bar. I had a favorable opportunity of seeing what the approach to the mouth of the river is like; the wind was dead against us, and we had to "beat" for several miles; the "lead" was kept going all the time, and I watched the soundings carefully. As soon as we reached three fathoms on one side of the channel, we turned and went in an almost opposite direction until we reached three fathoms on the other side. We continuously repeated this movement, and thus got a series of cross sections of the channel, and invariably in going from one three-fathoms point to the other the soundings in mid-channel showed five, six, seven fathoms, and, as we neared the mouth of the river, nine and ten fathoms.

Immediately inside the mouth of the river, which is nearly two miles wide, there is a large basin extending for several miles in each direction with a good depth of water; through this we passed and proceeded up the river to about 20 (twenty) miles from the entrance, where we anchored and went ashore in the ship's boat; it was here that I found the timber I have referred to.

The next day we proceeded in the ship's whaleboat about 35 or 40 (thirty-five or forty) miles up the river and camped for the night, returning the next day; from soundings we took on the way it was evident that the "Pearl" or a larger ship might have gone as far as we went.

The country appeared flat generally on both sides of the river, with one or two exceptions, and the soil rich—palms growing luxuriantly on either side; the water is perfectly fresh and sweet for miles before we turned back, but rising and falling several feet with the rise and fall of the tide.

We saw very few natives except at the mouth of the river; they appeared disposed to be friendly, and if kindly treated will probably make good allies; but those up the river are a strong muscular race, with finer limbs than any I have before seen; they look self-reliant and as if they had "rights;" they would probably be troublesome if harshly treated.

Captain Pennefather some months ago prepared a rough chart of what he had then seen of the river, and wherever he has recorded actual measurements, soundings, &c., the utmost reliance may be placed upon their accuracy; where he has had to exercise his judgment in estimating distances, I am satisfied he has kept well and safely within the mark.

I think it very desirable that none of the land on this river should be alienated until it is ascertained what will be required for supplying the wants of the railway. A further survey of the river is desirable, although I am satisfied, from what I have seen, that there would be no difficulty in getting the timber away. There are many points where convenient wharves or staging could be erected at very trifling cost.

I have, &c.,

ROBT. WATSON, C.E., M. INST. C.E.

The Honourable The Colonial Secretary.